SUTURING PRINCIPLES
AND TECHNIQUES IN LABORATORY ANIMAL SURGERY
MANUAL AND DVD

SUTURING PRINCIPLES
AND TECHNIQUES IN LABORATORY ANIMAL SURGERY

MANUAL AND DVD

John J. Bogdanske
Scott Hubbard-Van Stelle
Margaret Rankin Riley
Beth M. Schiffman

CRC Press
Taylor & Francis Group
Boca Raton   London   New York

CRC Press is an imprint of the
Taylor & Francis Group, an **informa** business

CRC Press
Taylor & Francis Group
6000 Broken Sound Parkway NW, Suite 300
Boca Raton, FL 33487-2742

© 2013 by Taylor & Francis Group, LLC
CRC Press is an imprint of Taylor & Francis Group, an Informa business

No claim to original U.S. Government works

Printed on acid-free paper
Version Date: 20130426

International Standard Book Number-13: 978-1-4665-5343-9 (DVD)

This book contains information obtained from authentic and highly regarded sources. Reasonable efforts have been made to publish reliable data and information, but the author and publisher cannot assume responsibility for the validity of all materials or the consequences of their use. The authors and publishers have attempted to trace the copyright holders of all material reproduced in this publication and apologize to copyright holders if permission to publish in this form has not been obtained. If any copyright material has not been acknowledged please write and let us know so we may rectify in any future reprint.

Except as permitted under U.S. Copyright Law, no part of this book may be reprinted, reproduced, transmitted, or utilized in any form by any electronic, mechanical, or other means, now known or hereafter invented, including photocopying, microfilming, and recording, or in any information storage or retrieval system, without written permission from the publishers.

For permission to photocopy or use material electronically from this work, please access www.copyright.com (http://www.copyright.com/) or contact the Copyright Clearance Center, Inc. (CCC), 222 Rosewood Drive, Danvers, MA 01923, 978-750-8400. CCC is a not-for-profit organization that provides licenses and registration for a variety of users. For organizations that have been granted a photocopy license by the CCC, a separate system of payment has been arranged.

Trademark Notice: Product or corporate names may be trademarks or registered trademarks, and are used only for identification and explanation without intent to infringe.

Library of Congress Cataloging-in-Publication Data

Suturing principles and techniques in laboratory animal surgery : manual and DVD / John J. Bogdanske ... [et al.].
  p. ; cm.
  Includes bibliographical references and index.
  Summary: "Regardless of how a wound occurs, it represents a disruption of normal tissue, and when a wound cannot heal by itself, a method must be employed to provide the wounded tissue with strength and stability until it heals. The most common methods used for this purpose are sutures, staples, and tissue glue. Whether you are an experienced surgeon, instructor, trainer, or beginning technician, this book can be used to introduce the materials and instruments required for suturing. These topics are covered: instrumentation; needle types and applications; suture material, packaging and applications. Additionally, various suture patterns and associated principles for proper knot tying are thoroughly covered, including step-by-step descriptions"--Provided by publisher.
  ISBN 978-1-4665-5343-9 (softcover : alk. paper)
  1. Laboratory animals--Surgery. 2. Laboratory animals--Diseases. 3. Veterinary traumatology. 4. Veterinary surgery. I. Bogdanske, John J.
  [DNLM: 1. Surgery, Veterinary--methods. 2. Suture Techniques--veterinary. 3. Animals, Laboratory--surgery. 4. Sutures--veterinary. 5. Wounds and Injuries--surgery. 6. Wounds and Injuries--veterinary. SF 914.3]
  SF996.5.S88 2013
  636.089'79178--dc23   2012048943

Visit the Taylor & Francis Web site at
http://www.taylorandfrancis.com

and the CRC Press Web site at
http://www.crcpress.com

Contents

disclaimers ix
before you proceed xi
    suture pattern description xi
    suture pattern narration xi
    suture pattern handout xii
welcome xiii
acknowledgment xv
introduction xvii
    suggested reading xviii

1   instrumentation 1

    introduction 1
    needle holder 1
    tissue forceps 4
    operating scissors 5
    scalpel handle 6
    blade 7
    work cited 8

2   needle types 9

    introduction 9
    needle components 10
        The Eye 10
        The Body 11
        The Point 11
    work cited 12
    suggested reading 12

3 suture material ... 13

suture selection ... 13
suture types ... 14
    Monofilament and Multifilament ... 14
    Absorbable and Nonabsorbable ... 14
the package ... 16
principles to remember when choosing suture ... 16
suggested reading ... 17

4 suture pattern descriptions and DVD narration ... 19

pattern: square knot/suture knot ... 19
    Principal Use ... 19
    Description ... 19
pattern: surgeon's knot ... 22
    Principal Use ... 22
    Description ... 22
pattern: simple interrupted ... 24
    Principal Use ... 24
    Description ... 24
pattern: simple continuous ... 27
    Principal Use ... 27
    Description ... 27
pattern: Ford interlocking or lockstitch ... 31
    Principal Use ... 31
    Description ... 31
pattern: cruciate ... 35
    Principal Use ... 35
    Description ... 35
pattern: interrupted horizontal mattress ... 38
    Principal Use ... 38
    Description ... 38
pattern: continuous horizontal mattress ... 41
    Principal Use ... 41
    Description ... 41
pattern: subcuticular ... 46
    Principal Use ... 46
    Description ... 46
pattern: skin staples ... 51
    Principal Use ... 51
    Description ... 51

pattern: skin staples—removal 53
  Principal Use 53
  Description 53
pattern or technique: suture ligation 54
  Principal Use 54
  Description 54

5 suture patterns: handouts 57

square knot/suture knot 57
  Principal use 57
  Description 57
  Procedure 58
surgeon's knot 60
  Principal Use 60
  Description 60
  Procedure 60
simple interrupted 63
  Principal Use 63
  Description 63
  Procedure 63
simple continuous 66
  Principal Use 66
  Description 66
  Procedure 66
Ford interlocking or lockstitch 70
  Principal Use 70
  Description 70
  Procedure 70
cruciate 74
  Principal Use 74
  Description 74
  Procedure 74
interrupted horizontal mattress 77
  Principal Use 77
  Description 77
  Procedure 77
continuous horizontal mattress 80
  Principal Use 80
  Description 80
  Procedure 81

subcuticular 84
    Principal Use 84
    Description 84
    Procedure 85

skin staples 91
    Principal Use 91
    Description 91
    Procedure 91

skin staples: removal 94
    Principal Use 94
    Description 94
    Procedure 94

6 study breaks: DVD narration 95

7 suture considerations for exotics and other species 97
  introduction 97
  rabbit 98
  ferret 98
  birds 98
  reptiles and amphibians 99
  fish 99
  works cited 99

glossary 101
index 103

disclaimers

This training DVD is not intended for use by individuals who lack thorough training and understanding of animal biosafety. This DVD should be used as a resource or refresher for techniques previously learned through animal biosafety training.

The techniques depicted in this DVD must be performed in compliance with institutionally approved animal protocols, observing all institutional and governmental safety regulations with regard to the use of appropriate safety equipment and procedures. If you lack a thorough understanding of the institutional and governmental rules applicable to the techniques depicted in this DVD, you must not perform these techniques.

The techniques depicted in this DVD must be performed only in an approved animal facility under the guidance of animal care specialists.

The techniques depicted in this DVD take place in a controlled laboratory setting and present a risk of serious bodily injury (including death) through the following exposure hazards: sharp objects, chemical or pharmaceutical agents, and biohazards, including but not limited to the bodily fluids and waste of mice or rats. These hazards may be avoided only by complying with a thorough laboratory safety program that complies with all applicable institutional and governmental rules. If your facility lacks such a program or if you do not thoroughly understand all aspects of that program, you must not perform the techniques depicted in this DVD.

before you proceed

To reach suturing nirvana, we recommend that the user take time to become familiar with the content of the first three sections of this book—"Instrumentation," "Needle Types," and "Suture Material." The user will acquire background information and learn basic suturing principles that will be helpful when reading the pattern descriptions and viewing the instructional DVD.

suture pattern description

Each suture pattern description includes the common name, principal use, and detailed step-by-step instructions on how to complete the pattern. The suture pattern description is more comprehensive than the suture pattern narration.

suture pattern narration

An audible narration accompanies the video clip for each suture pattern on the instructional DVD. A written copy of the narration is provided in the book along with each pattern description. The narration describes the suturing technique shown in the video demonstration and is intended for use only in conjunction with the instructional DVD.

suture pattern handout

The suture pattern handout includes the common name, principal use, brief pattern description, and sequential instructions for each pattern, accompanied by high-quality pictures.

welcome

The trainers at the Research Animal Resources Center (RARC), University of Wisconsin–Madison, want to thank you for taking the time to view this training DVD. The techniques demonstrated are those approved by the University of Wisconsin Institutional Animal Care and Use Committee (IACUC) and are commonly used on research animals. It is our hope that you find this information both educational and useful.

acknowledgment

We want to thank Jennifer Gaudio and Sarah Newman for their countless hours of editing, wordsmithing, and most of all patience throughout the compilation of this book. Your contributions to the Training Office, both personally and professionally, have been outstanding. Here's looking forward to many more endeavors together. On Wisconsin!

introduction

It has been documented that eyed suture needles were invented sometime between 50,000 and 30,000 BC, with bone needles becoming common by 20,000 BC. As surgery techniques advanced, so did the method of wound closure. Whether it was the Native Americans and their use of cautery, East African tribes ligating vessels with tendons, or South African tribes using large black ants, the goal was to help aid in the successful outcome of a surgical procedure.

Regardless of whether the wound is created by chance or during a surgical procedure, it represents a disruption of normal tissue. If the wound cannot heal by itself, a method must be employed to provide the wounded tissue with enough strength and stability until it heals. The most common methods used for this purpose are sutures, staples, and tissue glue.

Whether you are an experienced surgeon, instructor, trainer, or beginning technician, this book and DVD can be used to introduce you to the materials and instruments required for suturing. The following topics are covered: instrumentation; needle types and applications; and suture material, packaging, and applications. Last, various suture patterns and associated principles for proper knot tying are thoroughly covered, including step-by-step descriptions, video clips with narration, and handouts for each suture pattern.

Most of the demonstrations shown in the DVD are performed using a cadaver pig's foot, which is also used in the Research Animal Resources Center (RARC) Laboratory Animal Surgery class. The use of a pig's foot gives students the opportunity to practice suture patterns and to handle real tissue.

This book and DVD will help answer suture-related questions and aid the reader in making educated decisions about basic wound closure.

suggested reading

Edlich RF. *Surgical Knot Tying Manual*, 3rd ed. Covidien, Norwalk, CT, 2008.

Mackenzie, D. History of sutures, *Medical History* 17(2):158–168, 1973. ncbi.nlm.nih.gov

instrumentation

introduction

Knowing and understanding basic surgical principles can help a surgeon choose appropriate instrumentation for a specific procedure or surgery. Using the correct instrument will prevent tissue trauma while proper hand placement on the instrument allows for good stabilization and tissue manipulation. A needle holder, toothed tissue forceps, and scissors are instruments most commonly used when suturing. Needle holders and scissors are held at the level of the most distal knuckle of the thumb and ring finger. The surgeon can further increase the stability of the instrument by placing the index finger on or near the box lock while resting the middle finger on the finger loop above the ring finger (Figure 1.1).

When handling tissue, use only the tips of the instruments, taking care not to include any peripheral tissue when closing or locking an instrument. This can lead to trauma and tissue necrosis in unintended areas.

needle holder

There are various types of needle holders, all of which are used to pass a curved needle through tissue. In this book, we limit our discussion to the Olsen Hegar and the Mayo Hegar (Figure 1.2). Comparing the two, notice the only marked difference between them is the built-in scissors of the Olsen Hegar inidicated by the letter A

1

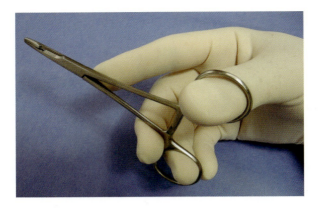

Fig. 1.1

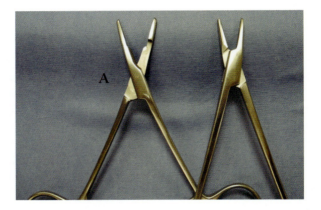

Fig. 1.2

in Figure 1.2. This feature allows the surgeon to cut suture material without switching between the needle holder and operating scissors, thus increasing efficiency.

Choose a needle holder that is the appropriate size and type for the animal, type of procedure, and needle size. A small needle should be held with a needle holder with small, fine jaws; needle holders with heavier jaws are required to properly hold larger needles.

The basic anatomy of the needle holder is shown in Figure 1.3.

Placement of the needle at the tip of the jaws is important for proper needle control. Grasp the needle two-thirds of the needle length from the tip and secure it in the needle holder by engaging the ratchet lock. If grasped correctly, the needle will be perpendicular (Figure 1.4) to the needle holder.

instrumentation 3

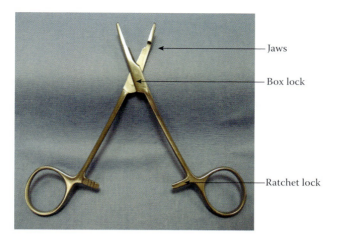

Fig. 1.3

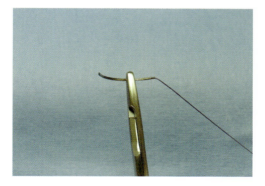

Fig. 1.4

Grasping the needle too close to the tip (Figure 1.5) will prevent the needle from being driven completely through the tissue. Grasping it too close to the swaged end (Figure 1.6) may result in the needle bending or even breaking. Check the alignment of the needle holder jaws regularly to ensure they have not twisted or turned over time and use.

Study Break 1

Watch the video, "Needle Placement on Needle Holder," for the proper way to place the needle on the needle holder.

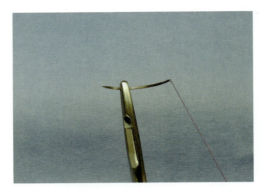

Fig. 1.5

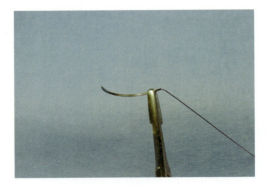

Fig. 1.6

tissue forceps

Typically, forceps are used to hold or pick up tissue that is too small for hands or fingers. In suturing, they are used to stabilize tissue during suture placement. This book focuses on three types of forceps: tissue forceps without teeth, which may be crosshatched or smooth; rat toothed, or Adson, forceps; and serrated, or Adson Brown, forceps. Figure 1.7 shows the three forceps and their distinct tips, which are from left to right tissue, Adson, and Adson Brown forceps.

Tissue forceps without teeth are primarily used to move dressings, remove sutures, or perform other similar procedures. Although counterintuitive, using forceps with serrations or teeth may cause less

Fig. 1.7

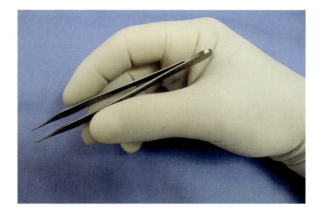

Fig. 1.8

tissue damage than smooth tissue forceps as less pressure is needed to grasp the tissue with toothed forceps. Therefore, Adson and Adson Brown forceps are more commonly used to grasp tissue when suturing.

Forceps are held between the thumb, index, and middle fingers as shown in Figure 1.8. This hand position provides control while limiting the amount of pressure that can be applied to the tissue.

operating scissors

Operating scissors are generally classified by their tips, such as blunt-blunt, blunt-sharp, or sharp-sharp (Figure 1.9). The Metzenbaum,

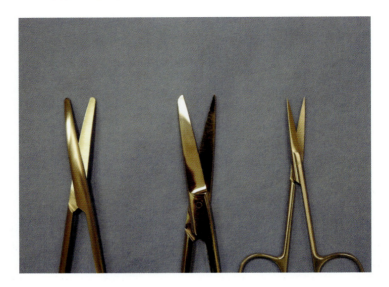

Fig. 1.9

Mayo, and Iris scissors (shown from left to right in Figure 1.9) are commonly used for surgical procedures. Most scissors are made of stainless steel, but some manufacturers place tungsten carbide along the cutting edge to increase the hardness. This hardness allows for a sharper cutting edge that results in smoother cuts.

Scissors are held at the level of the most distal knuckle of the thumb and ring finger. The surgeon can further increase stability by placing the index finger on or near the box lock while resting the middle finger on the finger loop above the ring finger. Cutting with scissors is similar to sharp dissection; cut using the tips of the scissor blades while ensuring the blades never completely close as they move through the tissue. Cutting too close to the fulcrum will weaken the cutting effect and potentially crush or damage the tissue.[1]

scalpel handle

The size of the scalpel handle is found on one side of the handle base and corresponds to blade sizes. In lab animal surgery, the most common handle sizes are the No. 3 and No. 4. Blade choices for the No. 3 handle include the Nos. 10, 11, 12, and 15, with the No. 10 and No. 15 commonly used in surgery (Figure 1.10).

instrumentation 7

Fig. 1.10

blade

Attach the blade onto the handle correctly to ensure safe and proper use. Once the blade packaging is open, observe that the number of the blade can be found on one side of the base. With the number facing up, grasp the blade along the back (opposite the cutting edge) and above the opening with a needle holder (Figure 1.11). Lock the ratchet lock, securing the blade on the needle holder. Tilt the blade slightly and slide it over the tip of the blade handle. Push the blade down until it snaps onto the handle by putting pressure on the box lock with the index finger.

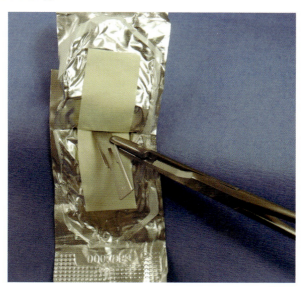

Fig. 1.11

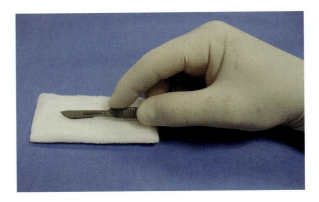

Fig. 1.12

Hold the scalpel handle between the fingertips and thumb, with the index finger resting on the spine of the blade handle (Figure 1.12). Apply precise pressure on the blade to cut through the tissue, using the belly of the blade to make the incision. While making the cut, apply tension to the surface on either side of the incision using the index finger and thumb of the nondominant hand.

Study Break 2

Watch the video, "Attach and Remove Blade Safely," for the proper way to place a scalpel blade on the blade handle.

Study Break 3

Watch the video, "Make an Incision," to see an example of making an incision.

work cited

1. Slatter DH. *Textbook of Small Animal Surgery*, Elsevier Science, Philadelphia, PA, 2003.

2

needle types

introduction

Before deciding on a needle type for a particular procedure or wound closure, take the time to consider needle characteristics. The best needles are typically made from high-quality stainless steel and are as slim as possible without compromising strength or integrity. The needle must be held stable in the tip of the needle holder without the possibility of rotating or slipping. The needle must be sharp enough to penetrate the tissue cleanly, allowing the suture material to be pulled through the tissue with minimum resistance and trauma.

Choosing the correct surgical needle will also depend on the tissue being sutured. The needle should alter the tissue as little as possible; the only purpose of the needle is to introduce the suture into the tissue for apposition.[1]

In laboratory animal surgery, there are typically two needle types used: the cutting and the taper point. A cutting needle has at least two cutting edges and is triangular in shape (Figure 2.1). Cutting needles can be further divided into two common types: conventional and reverse cutting. The conventional needle has three cutting edges that form a triangular cross section with the third cutting edge on the inner concave curve of the needle. The reverse cutting has the third cutting edge on the outer convex curve, which makes the needle stronger and could reduce the risk of cutting out the tissue being sutured.

10   *suturing principles and techniques in laboratory animal surgery*

Fig. 2.1   **Fig. 2.2**

Taper point needles (Figure 2.2) are used in tissues that are easy to penetrate or in areas that have nerve fibers or are highly vascular. As the taper point is advanced, it essentially stretches the tissue without cutting.

needle components

The Eye

Although surgical needles vary in their use or application, all needles have three components: the eye, the body, and the point.

The eye can be one of two types: a closed eye that needs to be threaded or a swaged eye. The closed eye (Figure 2.3) is similar in appearance to the eye found on a household sewing needle. The shape of the eye can be round, square, or oblong. A major disadvantage of the eyed needle is that it must be threaded, adding time to the procedure and forcing the surgeon to pull a double strand of suture material through the tissue. The double strand creates a larger hole and increases disruption of the tissue. In addition, repeated use of the eyed needle causes it to become dull, making it increasingly difficult to use. The use of eyed needles has decreased in recent years as swaged needles have become more prominent and their advantages more evident.

Swaged needles offer many advantages. The suture material is attached or crimped to the end of the needle, making it one continuous unit (Figure 2.4). Having the suture attached in this configuration allows for the needle and suture to be drawn through the tissue

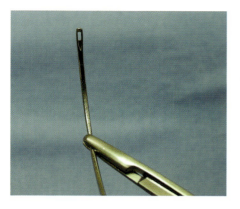 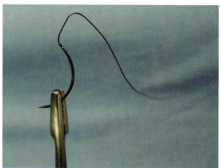

Fig. 2.3   Fig. 2.4

with minimal trauma. The swaged needle is intended for a single use, which eliminates problems such as dullness (cutting needles), damage, or contamination.

The Body

The body is the part of the needle that is grasped with the needle holder while suturing. To ensure minimal bleeding or leakage, the needle diameter needs to be similar in size to the suture material. The curvature of the needle can vary, lending itself to a variety of applications.

Straight needles are primarily used in areas that are easy to access and where fingers can do most of the manipulation, such as tendon and meniscus repairs and skin wound closures.

Curved needles are used more often than straight needles because the curve allows the surgeon to predict more precisely where the needle will pass through the tissue. Curvature can range from ¼ to ⅝ circle; therefore, this needle type can be used in a wide range of applications, from a superficial wound to a deep body cavity.

The Point

Although there are a number of specific needle points, this book addresses the conventional cutting, reverse cutting, and the taper point needles.

Cutting needles have at least two opposing cutting edges and are triangular in shape. Care must be taken when choosing a needle point to ensure tissue damage is kept to a minimum. The sharp edges of the cutting needle allow it to be passed easily through dense connective tissue or skin. The conventional cutting needle has a third cutting surface on the inside curve. Those using it for the first time need to be aware that it is possible to cut out or pull through the tissue toward the edge of the incision.

Reverse cutting needles have the cutting edge on the back of the curve rather than on the concave surface. This design offers advantages not found in the conventional cutting needle. First, the reverse cutting needle is stronger; second, the risk of tissue pull through or cutout is reduced. The reverse cutting needle is designed for tough tissue such as skin as well as more delicate tissue applications in ophthalmic and cosmetic surgery.

The taper point needle is sometimes called the round needle and is designed to penetrate and pass through the tissue without cutting it. Taper point needles are commonly used in easily penetrated tissue, such as subcutaneous layers, peritoneum, and abdominal viscera, as it minimizes the possibility of tearing through the fascia.

work cited

1. Trier WC. Considerations in the choice of surgical needles. *Surg Gynecol Obstet* 149:84, 1979.

suggested reading

Ethicon Wound Closure Manual. Ethicon Inc., Somerville, NJ. Internet Archive 2002.

suture material

suture selection

Suture has evolved over the years to include material designed specifically for surgical procedures. This specialized suture material decreases closure complications as well as postoperative infection.

Before choosing a specific suture, the surgeon needs to consider personal preference and the animal's physical factors. Personal preference may include the surgeon's area of expertise, wound-closing experience, and knowledge of wound healing. The animal's physical factors include the strength of the tissue to be sutured, the area in which the sutures will be placed, and the tissue's reaction to the material. Additional physical attributes to consider are the patient's age, weight, and health status.

The features of the suture material must also be considered. By understanding suture characteristics, such as size and strength, ease of use, and knot-holding ability, the surgeon is able to choose a material that minimizes tissue trauma yet retains its strength long enough for the wound or incision line to heal. Larger suture material will provide greater tensile strength but will also cause more tissue trauma when pulled through the tissue. The suture material should be as small as possible while providing enough support for the tissue to heal properly. Suture sizes range from 7 (large) to 11-0 (small). The following chart illustrates the sizing of suture material:

Largest/Strongest ⟶ **Smallest/Weakest**
7 6 5 4 3 2 1 0 2-0 3-0 4-0 5-0 6-0 7-0 8-0 9-0 10-0 11-0

suture types

Monofilament and Multifilament

Monofilament suture has one strand, whereas multifilament is comprised of several strands braided together. Each has benefits that make them appropriate for specific procedures.

Single-strand, monofilament suture passes through tissue with less resistance and is less likely to promote bacterial growth. Handling properties are good, and most surgeons find that knots are easily tied, but care must be taken in handling monofilament suture to prevent crimping. Any weak spot could potentially cause the suture to break, leading to incision failure.

The multiple strands of multifilament, or braided, suture will increase drag while passed through tissue, although the material is easier to work with and is stronger. Multifilament sutures have increased capillary activity and are not recommended for skin or areas that may be contaminated.

Absorbable and Nonabsorbable

Suture is further defined as absorbable or nonabsorbable. Both absorbable and nonabsorbable types can be either monofilament or multifilament.

Absorbable suture is designed to hold the wound together temporarily until it has healed enough to withstand normal tissue stresses. The suture is made from either collagen or a synthetic material and may absorb rapidly. It can also be treated chemically to slow the absorption rate. Absorbable suture material will lose most of its strength within a 60-day time frame and eventually be completely reabsorbed by the body. Advantages of absorbable suture are ease of use and decreased tissue reaction. However, the longevity of absorbable sutures is dependent on the patient's status. For instance, if the patient has a fever or infection and sutures are placed in an area that is fluid filled, the likelihood of the material being absorbed at a faster rate increases. This could cause premature suture failure, leading to postoperative complications. The amount of suture material used will also have an impact on the absorption rate. Be sure to use the appropriate suture pattern and suture material size. Avoid tying too many knots as this creates extra material for the body to break down.

Nonabsorbable sutures include, but are not limited to, stainless steel, silk, and nylon. The goal of nonabsorbable suture is to maintain its strength beyond 60 days, although the body may eventually absorb some materials over time. Nonabsorbable suture is primarily used in skin, where it will need to be removed, usually after 10–14 days, but may also be used inside the body, where it will be encapsulated in tissue. Table 3.1 provides specific information on common suture types.

TABLE 3.1: COMMON SUTURE

Material	Suture Type	Suture Name	Recommended for	Comments
Polydioxinone	A, mono	PDS[a]	Skin, percutaneous; interrupted pattern	Strong, long lasting
Gut	A, multi (natural)	Plain gut	Ligation; interrupted pattern	Weak, rapid absorption, high tissue reaction
Chromic gut	A, multi (natural)	Chromic gut	Ligation; interrupted pattern	Weak, rapid absorption, high tissue reaction
Polyglycolic	A, multi (braided)	DEXON[b]	Subcuticular, subcutaneous; continuous or interrupted pattern	Strength deteriorates quickly (~1 week)
Polyglactic	A, multi (braided)	VICRYL[a]	Subcuticular, subcutaneous; continuous or interrupted pattern	Commonly used
Nylon	N, mono (synthetic)	Ethilon[a] Dermalon[b]		Stiff and somewhat hard to work with
Polypropylene	N, mono (synthetic)	Prolene[a] Surgilene[b]	Percutaneous	Not the best choice for knot security; hard to work with
Silk	N, multi (natural)	Silk	Place in area for comfort; intraoral	Good knot security and easy to work with
Stainless steel	N, mono	Stainless steel	Skin, percutaneous	Typically used in nonhuman primates

Source: Modified from *Principles of Veterinary Suturing*, Marcel I. Perret-Gentil, DVM, MS, University Veterinarian and Director, Laboratory Animal Resources Center, The University of Texas at San Antonio.

Note: A, absorbable; N, nonabsorbable; percutaneous, in the skin; subcuticular, below the skin; mono, monofilament; subcutaneous, under the skin (e.g., fat layer); multi, multifilament.

[a] Ethicon.
[b] Sherwood Davis & Geck.

the package

The purpose of suture packaging is to protect the sterile contents and maintain the integrity of the needle and suture material prior to their use. The packaging provides the surgeon with information that will aid in choosing the appropriate suture material and needle type. Regardless of the manufacturer, basic information is typically listed as follows:

1. Size of the suture material. The sizing of suture material refers to the diameter.
2. Length of the suture material. This is listed in inches and centimeters.
3. Suture name and type. This will indicate a brand name and if the material is braided or a monofilament.
4. Needle type. The figure shows details for a reverse cutting needle.
5. Needle size. The picture shows the actual size of the needle. Also shown is a transverse slice of the needle to illustrate the needle point.

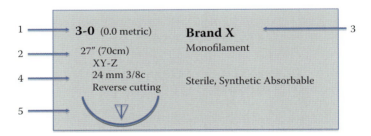

principles to remember when choosing suture

1. Tissue that heals slowly needs to be closed with nonabsorbable suture. The only exception would be an absorbable material that has been treated to decrease its absorption rate.
2. Tissue that heals quickly can be closed with absorbable suture.
3. Do not use multifilament (braided) suture in skin or contaminated wounds as it could lead to infection by wicking contaminants from the environment into the wound.

4. Absorbable monofilament sutures are a good choice for wounds that may be contaminated.
5. Choose the smallest suture size possible, keeping in mind the strength of the tissue to be sutured.

suggested reading

Ethicon Wound Closure Manual, Ethicon, Inc., Somerville, NJ. Internet Archive 2002.

suture pattern descriptions and DVD narration

Listed in this chapter are detailed descriptions and the text of the DVD narration for each suture pattern.

pattern: square knot/suture knot

Principal Use

A **square knot** is used to prohibit or eliminate slipping.
A **suture knot** is used to secure a knot in a suture pattern.

Description

The ability to tie a proper suture knot is essential for successfully completing any suture pattern. An improperly tied knot may result in what is called a *granny knot*. A granny knot can easily slip and release under tension and cause a suture pattern to come undone. Prior to tying a suture knot, it is necessary to understand how to tie a square knot correctly. A complete **square knot** equals 2 throws of suture. It takes a minimum of 2 throws to secure a knot. A **suture knot** is a square knot with the addition of at least 1 or 2 throws, equaling a minimum of 3 or 4 throws. More throws may be necessary depending on the suture material's coefficient of friction or if the wound is under tension.

To begin a suture knot, grasp the skin and evert the wound edge. Place the needle perpendicular to the skin and pass the needle through, following the curve of the needle. Grasp the skin on the opposite side, evert the edge, and place the needle perpendicular to the inside of the skin. Pass the needle through, following the curve of the needle. The suture should be equidistant from the wound edge on both sides of the incision. Pull the suture through, leaving a short end approximately 1 inch in length.

Place the needle holder between the two strands of suture. Wrap the long end of the suture around the tip of the needle holder once. Turn the needle holder toward the short end of the suture and grasp the short suture end. Pull the strands with equal tension in a horizontal motion, sliding the long end of the suture over the tip of the needle holder to the opposite side of the wound. The surgeon's hands will cross over each other as the strands are pulled; the short suture end will now be on the opposite side of the wound. Bring the wound edges together just until they touch, taking care not to overly tighten and evert the skin edges. This is the first **throw**.

For the second throw, place the needle holder between the two strands of suture and wrap the long end once around the tip of the needle holder. Turn the needle holder toward the short end of the suture and grasp the suture end. Pull the suture ends apart with equal tension in a horizontal motion, sliding the long end of the suture over the tip of the needle holder to the opposite side of the wound. Failure to alternate direction of the strands with each throw will result in a granny knot.

This is now a square knot with 2 throws of suture. To create a suture knot, an additional 1 or 2 throws must be added on top of the square knot, equaling a minimum of 3 or 4 throws. These throws will tighten and secure the knot. To create the third throw, repeat the steps previously described by placing the needle holder between the long and short strands of suture. Wrap the long end of suture once around the tip of the needle holder, grasp the short strand with the needle holder, and pull the suture over the tip of the needle holder to the opposite side of the wound. Repeat the process for the fourth and any additional throws, remembering to alternate the direction of the suture strands with each throw and to pull the strands with equal tension in a horizontal motion.

After placing the desired number of throws, cut both strands of suture above the knot, leaving the tabs long enough to grasp easily if the suture will be removed.

PATTERN NARRATION: SQUARE KNOT/SUTURE KNOT

Voice-Over: Tying a correct square knot is essential to successfully completing a suture knot.

To create a square knot, place the needle holder between the two strands of rope. Wrap the long strand around the needle holder; grasp the end of the short strand. Slide the long strand over the needle holder, bringing the short strand to the opposite side. This is the first throw.

Place the needle holder between the two strands, wrap the long strand around the needle holder, grasp the end of the short strand, and bring it back to the original side.

This is the second throw and a complete square knot.

An additional 1 or 2 throws, equaling a minimum of 3 or 4 throws, is required to complete a suture knot.

The following video demonstrates an entire suture knot.

Grasp the skin with a forceps, evert the wound edge, and place the needle perpendicular to the skin. Insert the needle through the skin on one side of the wound and exit the skin on the opposite side.

Grasp the needle with the needle holder and pull the suture material through the skin, leaving a short end approximately 1 inch in length.

Wrap the suture once around the needle holder, grasp the short end, and pull the suture strands horizontally to the opposite sides of the wound, bringing the wound edges together. This is the first throw.

For the second throw, place the needle holder between the strands, wrap the suture material once around the needle holder, grasp the short end, and pull the strands horizontally to the opposite sides of the wound.

Repeat the process for the third and any additional throws, remembering to alternate direction of the suture strands with each throw.

Cut both strands of suture above the knot, leaving the tabs long enough to grasp easily if the suture will be removed.

pattern: surgeon's knot

Principal Use

A surgeon's knot is used to maintain apposition of the wound edges, often in cases where the wound is under tension or a monofilament suture with a low coefficient of friction is used.

Description

A surgeon's knot is a binding knot used to prevent the first throw from becoming loose. The suture material is wrapped twice around the needle holder on the first throw of the surgeon's knot, creating increased frictional forces to help the suture stay in place. This is especially useful when using a monofilament that has a low coefficient of friction or if the wound is under tension.

To begin a surgeon's knot, grasp the skin and evert the wound edge. Place the needle perpendicular to the skin and pass the needle through, following the curve of the needle. Grasp the skin on the opposite side, evert the edge, and place the needle perpendicular to the inside of the skin. Pass the needle through, following the curve of the needle. The suture should be equidistant from the wound edge on both sides of the incision. Pull the suture through, leaving a short end that is approximately 1 inch in length.

Place the needle holder between the two strands of suture. Wrap the long end of the suture around the tip of the needle holder twice. Turn the needle holder toward the short end of the suture and grasp the short suture end. Pull the strands with equal tension in a horizontal motion, sliding the long end of the suture over the tip of the needle holder to the opposite side of the wound. The surgeon's hands will cross over each other as the strands are pulled; the short suture end will now be on the opposite side of the wound. Bring the wound edges together just until they touch, taking care not to evert the skin edges. This is the first throw.

For the second throw, place the needle holder between the two strands of suture and wrap the long end once around the tip of the needle holder. Turn the needle holder toward the short end of the suture and grasp the suture end. Pull the suture ends apart with equal tension in a horizontal motion, sliding the long end of the suture over the tip of the needle holder to the opposite side of the wound. Failure to alternate direction of the strands with each throw will result in a granny knot.

Each surgeon's knot is made up of 3 or 4 throws. To create the third throw, repeat the steps previously described by placing the needle holder between the long and short strands of suture. Wrap the long end of suture once around the tip of the needle holder, grasp the short strand with the needle holder, and pull the suture over the tip of the needle holder to the opposite side of the wound. Repeat the process for the fourth and any additional throws, remembering to alternate the direction of the suture strands with each throw and to pull the strands with equal tension in a horizontal motion. These additional throws will tighten and secure the knot.

After placing the desired number of throws, cut both strands of suture above the knot, leaving the tabs long enough to grasp easily if the suture will be removed.

PATTERN NARRATION: SURGEON'S KNOT

Voice Over: If there is increased tension on the incision or the suture material has a low coefficient of friction, a surgeon's knot can be used to prevent the first throw from becoming loose.

In this video demonstration, observe how the first throw relaxes and rises off the tissue after the suture material is wrapped once around the tip of the needle holder.

To better secure the first throw, wrap the suture material twice around the needle holder. The second wrap increases the contact area of the suture material and prevents slipping.

This double wrap is performed only on the first throw.

For the second throw, wrap the suture once around the needle holder, grasp the short end, and pull the strands horizontally to the opposite sides of the wound.

Repeat for the third and any additional throws, remembering to alternate direction of the suture strands.

Cut both strands of suture above the knot, leaving the tabs long enough to grasp easily if the suture will be removed.

A surgeon's knot consists of a minimum of 3 or 4 throws.

pattern: simple interrupted

Principal Use

This basic suture pattern is used to securely close a wound with accuracy of tissue apposition. It can be used in skin, muscle, organs, vessels, nerves, or fascia.

Description

The simple interrupted suture pattern is easy to place, has good tensile strength, and less potential for wound edema. This is a secure pattern that allows the surgeon to make adjustments as needed to align the wound edges while suturing.

The ability to tie a proper suture knot is essential for successfully completing the simple interrupted suture pattern. An improperly tied knot may result in what is called a granny knot. A granny knot can easily slip and release under tension and cause a suture pattern to come undone. The simple interrupted suture knot requires a minimum of 3 or 4 throws. More throws may be necessary depending on the suture material's coefficient of friction or if the wound is under tension.

To begin the simple interrupted suture pattern, grasp the skin and evert the wound edge. Place the needle perpendicular to the skin and pass the needle through, following the curve of the needle. Grasp the skin on the opposite side, evert the edge, and place the needle perpendicular to the inside of the skin. Pass the needle through, following the curve of the needle. The suture should be equidistant from the wound edge on both sides of the incision. Pull the suture through, leaving a short end that is approximately 1 inch in length.

Place the needle holder between the two strands of suture. Wrap the long end of the suture around the tip of the needle holder once. Turn the needle holder toward the short end of the suture and grasp the short suture end. Pull the strands with equal tension in a horizontal motion, sliding the long end of the suture over the tip of the needle holder to the opposite side of the wound. The surgeon's hands will cross over each other as the strands are pulled; the short suture end will now be on the opposite side of the wound. Bring the wound edges together just until they touch, taking care not to evert the skin edges. This is the first throw.

Note: If the wound is under tension or the suture material has a low coefficient of friction, such as a monofilament, a surgeon's knot may be preferred. Refer to the section "Surgeon's Knot."

For the second throw, place the needle holder between the two strands of suture and wrap the long end once around the tip of the needle holder. Turn the needle holder toward the short end of the suture and grasp the suture end. Pull the suture ends apart with equal tension in a horizontal motion, sliding the long end of the suture over the tip of the needle holder to the opposite side of the wound. Failure to alternate direction of the strands with each throw will result in a granny knot.

Each suture knot is made up of 3 or 4 throws. To create the third throw, repeat the steps previously described by placing the needle holder between the long and short strands of suture. Wrap the long end of suture once around the tip of the needle holder, grasp the short strand with the needle holder, and pull the suture over the tip of the needle holder to the opposite side of the wound. Repeat the process for the fourth and any additional throws, remembering to alternate the direction of the suture strands with each throw, and pull the strands with equal tension in a horizontal motion. These additional throws will tighten and secure the knot.

After placing the desired number of throws, cut both strands of suture above the knot, leaving the tabs long enough to grasp easily if the suture will be removed.

PATTERN NARRATION: SIMPLE INTERRUPTED

Voice Over: The simple interrupted suture pattern is a secure pattern that is easy to place. The ability to tie a suture knot is essential for completing this pattern.

Grasp the skin with a forceps, evert the wound edge, and place the needle perpendicular to the skin. Insert the needle through the skin on one side of the wound and exit the skin on the opposite side.

Grasp the needle with the needle holder and pull the suture material through the skin, leaving a short end approximately 1 inch in length.

continued

PATTERN NARRATION: SIMPLE INTERRUPTED
(continued)

A surgeon's knot is demonstrated here to secure the throw but may not always be necessary.

Wrap the suture twice around the needle holder, grasp the short end, and pull the suture strands horizontally to the opposite sides of the wound, bringing the wound edges together. This is the first throw.

For the second throw, place the needle holder between the strands, wrap the suture material once around the needle holder, grasp the short end, and pull the strands horizontally to the opposite sides of the wound.

For the third throw, wrap the suture material once around the needle holder, grasp the short end, and pull the strands horizontally, alternating direction of the suture strands.

Repeat the process for the fourth and any additional throws, remembering to alternate direction of the suture strands with each throw.

Cut both strands of suture material above the knot, leaving the tabs long enough to grasp easily if the suture will be removed.

Repeat this suture pattern the length of the incision, spacing the knots equidistantly.

pattern: simple continuous

Principal Use

The simple continuous pattern is useful when quick closure is desired, mainly in long wounds that are not under a great deal of tension and approximation of wound edges is acceptable. It may be advantageous in wounds requiring a more airtight or watertight closure.

Description

The simple continuous suture pattern is easy to place and allows for fast wound closure. The simple continuous uses less material, and fewer knots are tied with this pattern, generally resulting in less scarring. However, it offers less security because failure of either knot may result in failure of the entire suture pattern.

To begin the simple continuous suture pattern, grasp the skin and evert the wound edge. Place the needle perpendicular to the skin and pass the needle through, following the curve of the needle. Grasp the skin on the opposite side, evert the edge, and place the needle perpendicular to the inside of the skin. Pass the needle through, following the curve of the needle. The suture should be equidistant from the wound edge on both sides of the incision. Pull the suture through, leaving a short end that is approximately 1 inch in length.

Place the needle holder between the two strands of suture. Wrap the long end of the suture around the tip of the needle holder once. Turn the needle holder toward the short end of the suture and grasp the short suture end. Pull the strands with equal tension in a horizontal motion, sliding the long end of the suture over the tip of the needle holder to the opposite side of the wound. The surgeon's hands will cross over each other as the strands are pulled; the short suture end will now be on the opposite side of the wound. Bring the wound edges together just until they touch, taking care not to evert the skin edges. This is the first throw.

Note: If the wound is under tension or the suture material has a low coefficient of friction, such as a monofilament, a surgeon's knot may be preferred. Refer to the section, "Surgeon's Knot."

For the second throw, place the needle holder between the two strands of suture and wrap the long end once around the tip of the needle holder. Turn the needle holder toward the short end of the suture and grasp the suture end. Pull the suture ends apart with equal tension in a horizontal motion, sliding the long end of the suture over the tip of the needle holder to the opposite side of the wound. Failure to alternate direction of the strands with each throw will result in a granny knot.

Each suture knot is made up of 3 or 4 throws. To create the third throw, repeat the steps previously described by placing the needle holder between the long and short strands of suture. Wrap the long end of suture once around the tip of the needle holder, grasp the short strand with the needle holder, and pull the suture over the tip of the needle holder to the opposite side of the wound. Repeat the process for the fourth and any additional throws, remembering to alternate the direction of the suture strands with each throw, and pull the strands with equal tension in a horizontal motion. These additional throws will tighten and secure the knot.

After completing the suture knot with a minimum of 3 or 4 throws, leave the long end intact and cut only the short end, leaving a tab long enough to grasp if the suture will be removed. The long end of the suture strand will be used to create a running pattern the entire length of the incision.

Start the running pattern by grasping the skin below the initial knot and everting the wound edge. Place the needle perpendicular to the skin and pass the needle through, following the curve of the needle. Grasp the skin on the opposite side, evert the edge, and place the needle perpendicular to the inside of the skin. Pass the needle through, following the curve of the needle. The suture should be equidistant from the wound edge on both sides of the incision. Pull the suture through and repeat this step the entire length of the incision; bring the wound edges together just until they touch, taking care not to evert the skin edges.

Upon reaching the end of the incision, pull the suture strand through the skin, as previously described, until a loop is formed approximately 1 inch in length. This loop will represent the short end of the suture and will be used to tie the final suture knot.

Place the needle holder between the long strand of suture and the short loop. Wrap the long end of the suture once around the tip of the needle holder. Turn the needle holder toward the loop and grab the suture at the apex of the loop. Pull the strands with equal tension in a horizontal motion, sliding the long end of the suture over the tip of the needle holder to the opposite side of the wound. The surgeon's hands will cross over each other as the strands are pulled; the short loop will now be on the opposite side of the wound. Bring the wound edges together just until they touch, taking care not to evert the skin edges. This is the first throw.

Note: If the wound is under tension or the suture material has a low coefficient of friction, such as a monofilament, a surgeon's knot may be preferred. Refer to the section, "Surgeon's Knot."

For the second throw, place the needle holder between the long strand and short loop of suture and wrap the long end once around the tip of the needle holder. Turn the needle holder toward the loop and grab the suture at the apex of the loop. Pull the suture ends apart with equal tension in a horizontal motion, sliding the long end of the suture over the tip of the needle holder to the opposite side of the wound. Failure to alternate direction of the suture ends with each throw will result in a granny knot.

Each suture knot is made up of 3 or 4 throws. To create the third throw, repeat the steps previously described by placing the needle holder between the long strand and short loop of suture. Wrap the long end of suture once around the tip of the needle holder, grasp the apex of the loop with the needle holder, and pull the suture over the tip of the needle holder to the opposite side of the wound. Repeat the process for the fourth and any additional throws, remembering to alternate the direction of the suture strands with each throw, and pull the strands with equal tension in a horizontal motion. These additional throws will tighten and secure the knot.

Cut the loop and the long end of the suture above the knot, leaving 3 tabs long enough to grasp easily if the suture will be removed.

PATTERN NARRATION: SIMPLE CONTINUOUS

Voice Over: The simple continuous suture pattern allows for fast wound closure. However, failure of either knot may result in failure of the entire suture pattern.

Grasp the skin with a forceps, evert the wound edge, and place the needle perpendicular to the skin. Insert the needle through the skin on one side of the wound and exit the skin on the opposite side.

Grasp the needle with the needle holder and pull the suture material through the skin, leaving a short end approximately 1 inch in length.

continued

PATTERN NARRATION: SIMPLE CONTINUOUS
(continued)

A surgeon's knot is demonstrated here to secure the throw but may not always be necessary.

Wrap the suture twice around the needle holder, grasp the short end, and pull the suture strands horizontally to the opposite sides of the wound, bringing the wound edges together. This is the first throw.

For the second throw, wrap the suture material once around the needle holder, grasp the short end, and pull the strands horizontally to the opposite side of the wound.

Repeat the process for the third and any additional throws, remembering to alternate direction of the suture strands with each throw.

Cut only the short strand of suture material, leaving the tab long enough to grasp if the suture will be removed.

Grasp the skin below the knot, evert the wound edge, and place the needle perpendicular to the skin. Insert the needle through the skin on one side of the wound and exit the skin on the opposite side. The suture should be equidistant from the wound edge on both sides of the pattern. Pull the suture through and repeat this step the entire length of the incision.

To complete and secure the continuous suture pattern, pull the suture through the skin, leaving a loop, approximately 1 inch in length, to use as the short end.

Wrap the long end of the suture once around the needle holder, grasp the loop at the apex, and pull the suture strands horizontally to the opposite sides of the wound, bringing the wound edges together. This is the first throw.

For the second throw, wrap the suture material once around the needle holder, grasp the loop at the apex, and pull the loop to the opposite side of the wound.

Repeat the process for the third and any additional throws, remembering to alternate direction of the suture strands.

Cut the loop and the long end of the suture, leaving 3 tabs long enough to grasp easily if the suture will be removed.

pattern: Ford interlocking or lockstitch

Principal Use

The Ford interlocking or lockstitch is useful for closing long skin incisions or wounds under moderate tension.

Description

The lockstitch suture pattern allows for quick closure of a wound and is often more secure than the simple continuous pattern in the case of knot failure. The lockstitch requires more suture material and is more difficult to remove. It is most often used for skin closure in large animals.

To begin the lockstitch suture pattern, grasp the skin and evert the wound edge. Place the needle perpendicular to the skin and pass the needle through, following the curve of the needle. Grasp the skin on the opposite side, evert the edge, and place the needle perpendicular to the inside of the skin. Pass the needle through, following the curve of the needle. The suture should be equidistant from the wound edge on both sides of the incision. Pull the suture through, leaving a short end that is approximately 1 inch in length.

Place the needle holder between the two strands of suture. Wrap the long end of the suture around the tip of the needle holder once. Turn the needle holder toward the short end of the suture and grasp the short suture end. Pull the strands with equal tension in a horizontal motion, sliding the long end of the suture over the tip of the needle holder to the opposite side of the wound. The surgeon's hands will cross over each other as the strands are pulled; the short suture end will now be on the opposite side of the wound. Bring the wound edges together just until they touch, taking care not to evert the skin edges. This is the first throw.

Note: If the wound is under tension or the suture material has a low coefficient of friction, such as a monofilament, a surgeon's knot may be preferred. Refer to the section, "Surgeon's Knot."

For the second throw, place the needle holder between the two strands of suture and wrap the long end once around the tip of the needle holder. Turn the needle holder toward the short end of the suture and grasp the suture end. Pull the suture ends apart with equal tension in a horizontal motion, sliding the long end of the suture over the tip of the needle holder to the opposite side of the wound. Failure to alternate direction of the strands with each throw will result in a granny knot.

Each suture knot is made up of 3 or 4 throws. To create the third throw, repeat the steps previously described by placing the needle holder between the long strand and short strand of suture. Wrap the long end of suture once around the tip of the needle holder, grasp the short strand with the needle holder, and pull the suture over the tip of the needle holder to the opposite side of the wound. Repeat the process for the fourth and any additional throws, remembering to alternate the direction of the suture strands with each throw, and pull the strands with equal tension in a horizontal motion. These additional throws will tighten and secure the knot.

After completing the knot with a minimum of 3 or 4 throws, leave the long end intact and cut only the short end, leaving a tab long enough to grasp if the suture will be removed. The long end of the suture strand will be used to create a running pattern the entire length of the incision.

Start the running pattern by grasping the skin below the initial knot and everting the wound edge. Place the needle perpendicular to the skin and pass the needle through, following the curve of the needle. Grasp the skin on the opposite side, evert the edge, and place the needle perpendicular to the inside of the skin. Pass the needle through, following the curve of the needle. The suture should be equidistant from the wound edge on both sides of the incision.

Pass the needle through the loop of the preceding stitch. Pull the suture through the loop and repeat this step the entire length of the incision; bring the wound edges together just until they touch, taking care not to evert the skin edges.

Upon reaching the end of the incision, pull the suture through the skin as previously described until a loop is formed approximately 1 inch in length. This loop will represent the short end of the suture and will be used to tie the final suture knot.

Place the needle holder between the long strand of suture and the short loop. Wrap the long end of the suture once around the tip of the needle holder. Turn the needle holder toward the loop and grab the suture at the apex of the loop. Pull the strands with equal tension in a horizontal motion, sliding the long end of the suture over the tip of the needle holder to the opposite side of the wound. The surgeon's hands will cross over each other as the strands are pulled; the short loop will now be on the opposite side of the wound. Bring the wound edges together just until they touch, taking care not to evert the skin edges. This is the first throw.

Note: If the wound is under tension or the suture material has a low coefficient of friction, such as a monofilament, a surgeon's knot may be preferred. Refer to the section, "Surgeon's Knot."

For the second throw, place the needle holder between the long strand and short loop of suture and wrap the long end once around the tip of the needle holder. Turn the needle holder toward the loop and grab the suture at the apex of the loop. Pull the suture ends apart with equal tension in a horizontal motion, sliding the long end of the suture over the tip of the needle holder to the opposite side of the wound. Failure to alternate direction of the suture ends with each throw will result in a granny knot.

Each suture knot is made up of 3 or 4 throws. To create the third throw, repeat the steps previously described by placing the needle holder between the long strand and short loop of suture. Wrap the long end of suture once around the tip of the needle holder, grasp the apex of the loop with the needle holder, and pull the suture over the tip of the needle holder to the opposite side of the wound. Repeat the process for the fourth and any additional throws, remembering to alternate the direction of the suture strands with each throw, and pull the strands with equal tension in a horizontal motion. These additional throws will tighten and secure the knot.

Cut the loop and the long end of the suture above the knot, leaving 3 tabs long enough to grasp easily if the suture will be removed.

PATTERN NARRATION: FORD INTERLOCKING OR LOCKSTITCH

Voice Over: The Ford interlocking or lockstitch suture pattern allows for quick closure of a wound and is often more secure than the simple continuous pattern in the case of knot failure.

Grasp the skin with a forceps, evert the wound edge, and place the needle perpendicular to the skin. Insert the needle through the skin on one side of the wound and exit the skin on the opposite side.

Grasp the needle with the needle holder and pull the suture material through the skin, leaving a short end approximately 1 inch in length.

continued

PATTERN NARRATION: FORD INTERLOCKING OR LOCKSTITCH (continued)

A surgeon's knot is demonstrated here to secure the throw but may not always be necessary.

Wrap the suture twice around the needle holder, grasp the short end, and pull the suture strands horizontally to the opposite sides of the wound, bringing the wound edges together until they touch. This is the first throw.

For the second throw, wrap the suture material once around the needle holder, grasp the short end, and pull the strands horizontally, alternating direction of the strands.

Repeat the process for the third and any additional throws, remembering to alternate direction of the suture strands with each throw.

Cut only the short strand of suture material, leaving the tab long enough to grasp if the suture will be removed.

Grasp the skin below the knot, evert the wound edge, and place the needle perpendicular to the skin. Insert the needle through the skin on one side of the wound and exit the skin on the opposite side. The suture should be equidistant from the wound edge on both sides of the pattern.

Pass the needle through the loop of the preceding stitch.

Pull the suture through the loop, bringing the wound edges together until they touch. Repeat this step the entire length of the incision.

To complete and secure the lockstitch suture pattern, pull the suture through the skin, pass the needle through the loop of the preceding stitch, leaving a loop approximately 1 inch in length to use as the short end.

Wrap the suture once around the needle holder, grasp the loop at the apex, and pull the suture strands to the opposite sides of the wound.

For the second throw, wrap the suture material once around the needle holder, grasp the loop at the apex, and bring the loop to the opposite side of the wound.

Repeat the process for the third and any additional throws, remembering to alternate direction of the suture strands.

Cut the loop and the long end of the suture, leaving 3 tabs long enough to grasp easily if the suture will be removed.

pattern: cruciate

Principal Use

The cruciate pattern is primarily used for closing skin wounds under tension.

Description

The cruciate suture pattern is quick and easy to place. It crosses over itself, allowing for a strong closure that is ideal for skin wounds under tension.

To begin the cruciate suture pattern, grasp the skin and evert the wound edge. Place the needle perpendicular to the skin and pass the needle through, following the curve of the needle. Grasp the skin on the opposite side, evert the edge, and place the needle perpendicular to the inside of the skin. Pass the needle through, following the curve of the needle. The suture should be equidistant from the wound edge on both sides of the incision. Pull the suture through, leaving a short end that is approximately 1 inch in length.

Bring the suture across the wound, grasp and evert the skin below the initial needle insertion, and place the needle perpendicular to the skin. Pass the needle through the skin, following the curve of the needle. Grasp the skin on the opposite side, evert the edge, and place the needle perpendicular to the inside of the skin. Pass the needle through, following the curve of the needle. Pull the suture through until the wound edges come together, taking care not to invert the edges. This will form a diagonal pattern.

Place the needle holder between the two strands of suture. Wrap the long end of the suture around the tip of the needle holder once. Turn the needle holder toward the short end of the suture and grasp the short suture end. Pull the strands with equal tension in a horizontal motion, sliding the long end of the suture over the tip of the needle holder to the opposite side of the wound. The surgeon's hands will cross over each other as the strands are pulled; the short suture end will now be on the opposite side of the wound. Bring the wound edges together just until they touch, taking care not to invert the skin edges. This is the first throw.

Note: If the wound is under tension or the suture material has a low coefficient of friction, such as a monofilament, a surgeon's knot may be preferred. Refer to the section, "Surgeon's Knot."

For the second throw, place the needle holder between the two strands of suture and wrap the long end once around the tip of the needle holder. Turn the needle holder toward the short end of the suture and grasp the suture end. Pull the suture ends apart with equal tension in a horizontal motion, sliding the long end of the suture over the tip of the needle holder to the opposite side of the wound. Failure to alternate direction of the strands with each throw will result in a granny knot.

Each suture knot is made up of 3 or 4 throws. To create the third throw, repeat the steps previously described by placing the needle holder between the long and short strands of suture. Wrap the long end of suture once around the tip of the needle holder, grasp the short strand with the needle holder, and pull the suture over the tip of the needle holder to the opposite side of the wound. Repeat the process for the fourth and any additional throws, remembering to alternate the direction of the suture strands with each throw, and pull the strands with equal tension in a horizontal motion. These additional throws will tighten and secure the knot.

After placing the desired number of throws, cut both strands of suture above the knot, leaving the tabs long enough to grasp easily if the suture will be removed.

PATTERN NARRATION: CRUCIATE

Voice Over: The cruciate suture pattern is quick and easy to place. It crosses over itself, allowing for a strong closure that is ideal for skin wounds under tension.

Grasp the skin with a forceps, evert the wound edge, and place the needle perpendicular to the skin. Insert the needle through the skin on one side of the wound and exit the skin on the opposite side.

Grasp the needle with the needle holder and pull the suture material through the skin, leaving a short end approximately 1 inch in length.

Grasp and evert the skin below the initial needle insertion.

Insert the needle through the skin and exit the skin on the opposite side of the wound. Pull the suture through until the wound edges come together, taking care not to invert the edges.

A surgeon's knot is demonstrated here to secure the throw but may not always be necessary.

Wrap the suture twice around the needle holder, grasp the short end, and pull the suture strands horizontally to the opposite sides of the wound, bringing the wound edges together. This is the first throw.

For the second throw, wrap the suture once around the needle holder, grasp the short end, and pull the strands horizontally to the opposite sides of the wound.

Repeat the process for the third and any additional throws, remembering to alternate direction of the suture strands with each throw.

Cut both strands of suture material above the knot, leaving the tabs long enough to grasp easily if the suture will be removed.

Repeat this suture pattern the length of the incision, spacing the knots equidistantly.

pattern: interrupted horizontal mattress

Principal Use

The interrupted horizontal mattress is used for skin, muscle, or tendons that are under high tension.

Description

The interrupted horizontal mattress suture pattern provides good strength and wound eversion. It is generally used in areas that are under high tension. Less scarring occurs with this pattern because fewer knots are tied, although the number of needle insertions remains the same. The interrupted horizontal mattress also functions as a "stay stitch" to temporarily approximate the wound edge while allowing placement of a simple interrupted or simple continuous pattern.

To begin the interrupted horizontal mattress suture pattern, grasp the skin and evert the wound edge. Place the needle perpendicular to the skin and pass the needle through, following the curve of the needle. Grasp the skin on the opposite side, evert the edge, and place the needle perpendicular to the inside of the skin. Pass the needle through, following the curve of the needle. The suture should be equidistant from the wound edge on both sides of the incision. Pull the suture through, leaving a short end that is approximately 1 inch in length.

On the same side of the incision, stabilize the skin with a forceps below the location where the needle just exited. Place the needle perpendicular to the skin, insert through, and exit perpendicular to the skin on the opposite side of the wound. The suture should be equidistant from the wound edge on both sides of the incision. Grasp the needle with the needle holder and pull the suture material through the skin. Both the long and short ends of suture will now be on the same side of the wound.

Place the needle holder between the two strands of suture. Wrap the long end of the suture around the tip of the needle holder once. Turn the needle holder toward the short end of the suture and grasp the short suture end. Pull the strands with equal tension in a horizontal motion; the surgeon's hands will cross over each other as the strands are pulled. The long end of the suture will slide over the tip of the needle holder, and the short strand will be pulled to the opposite side. Bring the wound edges together just until they touch, taking care not to invert the skin edges. This is the first throw.

Note: If the wound is under tension or the suture material has a low coefficient of friction, such as a monofilament, a surgeon's knot may be preferred. Refer to the section, "Surgeon's Knot."

For the second throw, place the needle holder between the two strands of suture and wrap the long end once around the tip of the needle holder. Turn the needle holder toward the short end of the suture and grasp the suture end. Pull the suture ends apart with equal tension in a horizontal motion, sliding the long end of the suture over the tip of the needle holder to the opposite side. Failure to alternate direction of the strands with each throw will result in a granny knot.

Each suture knot is made up of 3 or 4 throws. To create the third throw, repeat the steps previously described by placing the needle holder between the long and short strands of suture. Wrap the long end of suture once around the tip of the needle holder, grasp the short strand with the needle holder, and pull the suture over the tip of the needle holder to the opposite side. Repeat the process for the fourth and any additional throws, remembering to alternate the direction of the suture strands with each throw, and pull the strands with equal tension in a horizontal motion. These additional throws will tighten and secure the knot.

After placing the desired number of throws, cut both strands of suture above the knot, leaving the tabs long enough to grasp easily if the suture will be removed.

PATTERN NARRATION: INTERRUPTED HORIZONTAL MATTRESS

Voice Over: The interrupted horizontal mattress suture pattern provides good strength and wound eversion. It is generally used in areas under high tension.

Grasp the skin with a forceps, evert the wound edge, and place the needle perpendicular to the skin. Insert the needle through the skin on one side of the wound and exit the skin on the opposite side.

continued

PATTERN NARRATION: INTERRUPTED
HORIZONTAL MATTRESS (continued)

Grasp the needle with the needle holder and pull the suture material through the skin, leaving a short end approximately 1 inch in length.

Stabilize the skin with a forceps below the location where the needle just exited. Place the needle perpendicular to the skin; insert through the skin and exit perpendicular to the skin on the opposite side of the wound.

Grasp the needle and pull the suture through the skin. Both the long and short ends of suture will now be on the same side of the wound.

A surgeon's knot is demonstrated here to secure the throw but may not always be necessary.

Place the needle holder between the two strands of suture. Wrap the suture twice around the needle holder, grasp the short end, and pull the suture strands horizontally to the opposite sides, bringing the wound edges together. This is the first throw.

For the second throw, wrap the suture once around the needle holder, grasp the short end, and pull the strands horizontally to the opposite sides.

Repeat the process for the third and any additional throws, remembering to alternate direction of the suture strands with each throw.

Cut both strands of suture material above the knot, leaving the tabs long enough to grasp easily if the suture will be removed.

Repeat this suture pattern the length of the incision, spacing the knots equidistantly.

pattern: continuous horizontal mattress

Principal Use

The continuous horizontal mattress is used for skin, muscle, or tendons that are under high tension.

Description

The continuous horizontal mattress suture pattern provides good strength and wound eversion. It is generally used in areas that are under high tension. Less scarring occurs with this pattern because fewer knots are tied, although the number of needle insertions remains the same. The continuous horizontal mattress suture pattern allows for faster wound closure than the interrupted horizontal mattress suture pattern; however, failure of either knot may result in failure of the entire suture pattern.

To begin the continuous horizontal mattress suture pattern, grasp the skin and evert the wound edge. Place the needle perpendicular to the skin and pass the needle through, following the curve of the needle. Grasp the skin on the opposite side, evert the edge, and place the needle perpendicular to the inside of the skin. Pass the needle through, following the curve of the needle. The suture should be equidistant from the wound edge on both sides of the incision. Pull the suture through, leaving a short end that is approximately 1 inch in length.

Place the needle holder between the two strands of suture. Wrap the long end of the suture around the tip of the needle holder once. Turn the needle holder toward the short end of the suture and grasp the short suture end. Pull the strands with equal tension in a horizontal motion, sliding the long end of the suture over the tip of the needle holder to the opposite side of the wound. The surgeon's hands will cross over each other as the strands are pulled; the short suture end will now be on the opposite side of the wound. Bring the wound edges together just until they touch, taking care not to evert the skin edges. This is the first throw.

Note: If the wound is under tension or the suture material has a low coefficient of friction, such as a monofilament, a surgeon's knot may be preferred. Refer to the section, "Surgeon's Knot."

For the second throw, place the needle holder between the two strands of suture and wrap the long end once around the tip of the needle holder. Turn the needle holder toward the short end of the suture and grasp the suture end. Pull the suture ends apart with equal tension in a horizontal motion, sliding the long end of the suture over the tip of the needle holder to the opposite side of the wound. Failure to alternate direction of the strands with each throw will result in a granny knot.

Each suture knot is made up of 3 or 4 throws. To create the third throw, repeat the steps previously described by placing the needle holder between the long and short strands of suture. Wrap the long end of suture once around the tip of the needle holder, grasp the short strand with the needle holder, and pull the suture over the tip of the needle holder to the opposite side of the wound. Repeat the process for the fourth and any additional throws, remembering to alternate the direction of the suture strands with each throw, and pull the strands with equal tension in a horizontal motion. These additional throws will tighten and secure the knot.

After completing the suture knot with a minimum of 3 or 4 throws, leave the long end intact and cut only the short end, leaving a tab long enough to grasp if the suture will be removed. The long end of the suture strand will then be used to create a running horizontal mattress pattern the entire length of the incision.

Start the running pattern by grasping the skin below the initial knot and everting the wound edge. Place the needle perpendicular to the skin and pass the needle through, following the curve of the needle. Grasp the skin on the opposite side, evert the edge, and place the needle perpendicular to the inside of the skin. Pass the needle through, following the curve of the needle, pulling the suture through the skin until the edges touch. The suture should be equidistant from the wound edge on both sides of the incision.

On the same side of the incision, stabilize the skin with a forceps below the location where the needle just exited. Place the needle perpendicular to the skin, insert the needle through, and exit perpendicular to the skin on the opposite side of the wound. The suture should be equidistant from the wound edge on both sides of the incision. Grasp the needle with the needle holder and pull the suture material through the skin. The long end of suture will now be on the same side of the wound as the initial knot. Pull the suture through and repeat these steps, alternating sides from right to left, then left to right, the entire length of the incision.

On reaching the end of the incision, pull the suture strand through the skin until a loop is formed approximately 1 inch in length. This loop will represent the short end of the suture and will be used to tie the final suture knot.

Place the needle holder between the long strand of suture and the short loop. Wrap the long end of the suture once around the tip of the needle holder. Turn the needle holder toward the loop and grab the suture at the apex of the loop. Pull the strands with equal tension in a horizontal motion, sliding the long end of the suture over the tip of the needle holder to the opposite side of the wound. The surgeon's hands will cross over each other as the strands are pulled; the short loop will now be on the opposite side of the wound. Bring the wound edges together just until they touch, taking care not to evert the skin edges. This is the first throw.

Note: If the wound is under tension or the suture material has a low coefficient of friction, such as a monofilament, a surgeon's knot may be preferred. Refer to the section, "Surgeon's Knot."

For the second throw, place the needle holder between the long strand and short loop of suture and wrap the long end once around the tip of the needle holder. Turn the needle holder toward the loop and grab the suture at the apex of the loop. Pull the suture ends apart with equal tension in a horizontal motion, sliding the long end of the suture over the tip of the needle holder to the opposite side of the wound. Failure to alternate direction of the suture ends with each throw will result in a granny knot.

Each suture knot is made up of 3 or 4 throws. To create the third throw, repeat the steps previously described by placing the needle holder between the long strand and short loop of suture. Wrap the long end of suture once around the tip of the needle holder, grasp the apex of the loop with the needle holder, and pull the suture over the tip of the needle holder to the opposite side of the wound. Repeat the process for the fourth and any additional throws, remembering to alternate the direction of the suture strands with each throw, and pull the strands with equal tension in a horizontal motion. These additional throws will tighten and secure the knot.

Cut the loop and the long end of the suture, leaving 3 tabs that will be easy to grasp if the suture will be removed.

PATTERN NARRATION: CONTINUOUS HORIZONTAL MATTRESS

Voice Over: The continuous horizontal mattress suture pattern provides good strength and wound eversion. It is generally used in areas that are under high tension. However, failure of either knot may result in failure of the entire suture pattern.

Grasp the skin with a forceps, evert the wound edge, and place the needle perpendicular to the skin. Insert the needle through the skin on one side of the wound and exit the skin on the opposite side.

Grasp the needle with the needle holder and pull the suture material through the skin, leaving a short end approximately 1 inch in length.

A surgeon's knot is demonstrated here to secure the throw but may not always be necessary.

Wrap the suture twice around the needle holder, grasp the short end, and pull the suture strands horizontally to the opposite sides of the wound, bringing the wound edges together. This is the first throw.

For the second throw, wrap the suture material once around the needle holder, grasp the short end, and pull the strands horizontally to the opposite sides of the wound.

Repeat the process for the third and any additional throws, remembering to alternate direction of the suture strands with each throw.

Cut only the short strand of suture material, leaving the tab long enough to grasp if the suture will be removed.

Grasp the skin below the knot, evert the wound edge, and place the needle perpendicular to the skin. Insert the needle through the skin on one side of the wound and exit the skin on the opposite side. The suture should be equidistant from the wound edge on both sides of the pattern.

Grasp the skin below the area the needle just exited. Insert the needle through and exit perpendicular to the skin on the opposite side of the wound.

Repeat these steps, alternating sides from left to right, right to left, the entire length of the incision.

To complete and secure the continuous horizontal mattress suture pattern, pull the suture through the skin, leaving a loop approximately 1 inch in length to use as the short end.

A surgeon's knot is again demonstrated here to secure the throw but may not always be necessary.

Wrap the long end of the suture twice around the needle holder, grasp the loop at the apex, and pull the suture strands horizontally to the opposite sides of the wound, bringing the wound edges together. This is the first throw.

For the second throw, wrap the suture material once around the needle holder, grasp the loop at the apex, and pull the hands apart, alternating direction of the strands.

Repeat the process for the third and any additional throws, remembering to alternate direction of the suture strands.

Cut the loop and remaining long end of suture material, leaving tabs long enough to grasp, making their removal easy.

pattern: subcuticular

Principal Use

The subcuticular suture pattern is used on areas where a visible suture line is not desirable. This pattern is advantageous when working with species that have a tendency to pick at exposed suture material, such as primates.

Description

The subcuticular suture pattern can be difficult to place, but when applied properly provides good skin apposition with little to no scarring. The entire pattern is applied into the dermal or subcuticular layer, just below the epidermis, allowing it to be hidden beneath the skin. This pattern can be helpful when working with species that have a tendency to pick at exposed suture material, such as primates. The subcuticular pattern is weaker than skin sutures; therefore, some skin closures may be added if desired.

A subcuticular suture pattern begins with an anchor knot buried beneath the skin. Starting at one end of the incision, insert the needle below the epidermis, deep into the dermal or subcuticular layer. Exit vertically through the dermal layer, just below the epidermis. Pull the suture through, leaving a short end that is approximately 1 to 2 inches in length. Then, on the opposite side of the incision, insert the needle into the dermal layer just below the epidermis and exit vertically out the deeper dermal or subcuticular layer.

Place the needle holder between the two strands of suture. Wrap the long end of the suture around the tip of the needle holder once. Turn the needle holder toward the short end of the suture and grasp the short suture end. Pull the strands with equal tension in a horizontal motion, sliding the long end of the suture over the tip of the needle holder to the opposite side of the wound. The short suture end will now be on the opposite side of the wound. This is the first throw.

For the second throw, place the needle holder between the two strands of suture and wrap the long end once around the tip of the needle holder. Turn the needle holder toward the short end of the suture and grasp the suture end. Pull the suture ends apart with equal tension in a horizontal motion, sliding the long end of the suture over the tip of the needle holder to the opposite side of the wound. Failure to alternate direction of the strands with each throw will result in a granny knot.

Each suture knot is made up of 3 or 4 throws. To create the third throw, repeat the steps previously described by placing the needle holder between the long and short strands of suture. Wrap the long end of suture once around the tip of the needle holder, grasp the short strand with the needle holder, and pull the suture over the tip of the needle holder to the opposite side of the wound. Repeat the process for the fourth and any additional throws, remembering to alternate the direction of the suture strands with each throw, and pull the strands with equal tension in a horizontal motion. These additional throws will tighten and secure the knot. The knot will be deep into the dermal or subcuticular layer, which results in a "buried knot."

After completing the anchor knot with a minimum of 3 or 4 throws, leave the long end intact and cut only the short end, just above the knot. The long end of the suture strand will be used to create a horizontal running pattern the entire length of the incision.

Start the running pattern by grasping the skin at the beginning of the incision and everting the wound edge. Place the needle under the epidermis into the dermal layer. Pass the needle horizontally through dermis. It is important to stay below the epidermal layer. Bring the wound edges together just until they touch, taking care not to invert the skin edges. Grasp the skin on the opposite side, evert the edge, and again place the needle under the epidermis into the dermal layer, and pass the needle horizontally through dermis, bringing the wound edges together. Continue passing the needle horizontally through the dermal layer, alternating sides the entire length of the incision.

On reaching the end of the incision, complete the suture pattern with a buried knot. Grasp the skin toward the end of the incision and evert the wound edge. Insert the needle below the epidermis, deep into the dermal or subcuticular layer. Exit vertically through the dermal layer, just below the epidermis. Pull the suture through until a loop is formed approximately 1 inch in length. This loop will represent the short end of the suture and will be used to tie the final suture knot. Then, on the opposite side of the incision, insert the needle vertically into the dermal layer, just below the epidermis, and exit vertically out the deeper dermal or subcuticular layer.

Place the needle holder between the long strand of suture and the short loop. Wrap the long end of the suture once around the tip of the needle holder. Turn the needle holder toward the loop and grab the suture at the apex of the loop. Pull the strands with equal tension in a horizontal motion, sliding the long end of the suture over the tip of the needle holder to the opposite side of the wound. The short loop will now be on the opposite side of the wound. This is the first throw.

For the second throw, place the needle holder between the long strand and short loop of suture and wrap the long end once around the tip of the needle holder. Turn the needle holder toward the loop and grab the suture at the apex of the loop. Pull the suture ends apart with equal tension in a horizontal motion, sliding the long end of the suture over the tip of the needle holder to the opposite side of the wound. Failure to alternate direction of the suture ends with each throw will result in a granny knot.

Each suture knot is made up of 3 or 4 throws. To create the third throw, repeat the steps previously described by placing the needle holder between the long strand and short loop of suture. Wrap the long end of suture once around the tip of the needle holder, grasp the apex of the loop with the needle holder and pull the suture over the tip of the needle holder to the opposite side of the wound. Repeat the process for the fourth and any additional throws, remembering to alternate the direction of the suture strands with each throw, and pull the strands with equal tension in a horizontal motion. These additional throws will tighten and secure the knot.

After completing the knot with a minimum of 3 or 4 throws, leave the long end intact and cut only the short loop, just above the knot.

Insert the needle vertically back into the incision, just behind the final buried knot. Turn the needle away from the incision line and exit the skin. Apply tension to the suture, pulling the buried knot deeper into the tissue. Cut the suture strand flush with the skin layer.

PATTERN NARRATION: SUBCUTICULAR

Voice Over: The subcuticular suture pattern is used on areas where a visible suture line is not desirable. This pattern is advantageous when working with species that have a tendency to pick at exposed suture material, such as primates.

The subcuticular begins with an anchor knot buried beneath the skin.

Grasp the skin with a forceps, evert the wound edge, and insert the needle deep into the dermal or subcuticular layer.

Exit vertically through the dermal layer, just below the epidermis. Pull the suture through; leaving a short end that is approximately 1 to 2 inches in length.

Insert the needle on the opposite side of the incision, just below the epidermis, and exit vertically out the deeper dermal or subcuticular layer.

Place the needle holder between the two strands, wrap the long end around the tip, grasp the short end, and pull both strands horizontally to the opposite sides of the wound, bringing the wound edges together. This is the first throw.

For the second throw, wrap the suture once around the needle holder, grasp the short end, and pull the strands to the opposite sides of the wound.

Repeat the process for the third and any additional throws, remembering to alternate direction of the suture strands with each throw.

Cut only the short strand of suture close to the knot.

Evert the wound edge, insert the needle horizontally above the knot below the epidermal layer, and pass it through the dermis. Pull the suture material through.

Repeat the placement of the needle under the epidermal layer on the opposite side of the wound, advancing the needle horizontally through the dermal layer.

Continue passing the needle through the dermal layer horizontally, under the epidermal layer, alternating sides the entire length of the incision.

The suture pattern will be completed with a buried knot. Grasp the skin, evert the wound edge, and insert the needle deep into the dermal or subcuticular layer. Exit vertically through the dermal layer just below the epidermis. Pull the suture through, leaving a loop approximately 1 inch in length.

Insert the needle through the dermal layer on the opposite side of the incision, just below the epidermis, and exit vertically out the deeper dermal or subcuticular layer.

Wrap the long end once around the needle holder, grasp the loop at the apex, and pull the suture strands horizontally to the opposite sides of the wound, bringing the wound edges together. This is the first throw.

continued

PATTERN NARRATION: SUBCUTICULAR
(continued)

For the second throw, place the needle holder between the strands, wrap the suture once around the needle holder, grasp the loop at the apex, and bring the loop to the opposite side of the wound.

Repeat the process for the third and any additional throws, remembering to alternate direction of the suture strands with each throw.

Cut the loop as close to the knot as possible.

Insert the needle vertically back into the incision, just behind the final buried knot. Turn the needle away from the incision line and exit the skin.

Apply tension to the suture, pulling the buried knot deeper into the tissue. Cut the suture strand flush with the skin layer.

pattern: skin staples

Principal Use

Skin staples are used to close skin wounds under tension. They may be used as an alternative to suture to prevent an animal from chewing through the suture line.

Description

Skin staples are quick and easy to apply and remove when closing skin wounds under tension. They are often used instead of suture to prevent an animal from chewing through the suture line and opening the wound.

To apply skin staples, grasp the skin on both sides of the incision with a forceps and evert the skin layers. Keeping the wound edges together, place the tip of the stapler against the everted skin, just below the forceps. Apply the staple by pressing the applicator together, causing the staple to insert into the skin layer. The staples should be placed equidistant from the wound edge on both sides of the incision.

Depending on the type of staple and applicator used, the staple may not be flush with the skin and may stand up off the skin layer. This allows room for swelling, which is often associated with tissue trauma.

Continue placing the staples, as described, along the length of the incision.

Staples are easily removed with a specified staple remover 10–14 days after application.

PATTERN NARRATION: SKIN STAPLES

Voice Over: Staples are commonly used to close wounds under tension. They may be used to ensure the incision remains intact in areas of the body where animals are known to chew through the suture material.

Grasp equal amounts of skin on both sides of the incision with a forceps and evert the wound edges.

Place the stapler against the everted skin, just below the forceps.

Squeeze the applicator between the thumb and index finger, inserting the staple into the skin.

The staples will stand up off the skin layer. This will allow room for swelling, which is often associated with tissue trauma.

Continue to place the staples equidistantly along the length of the incision.

pattern: skin staples—removal

Principal Use

A staple remover is used to remove skin staples from an incision 10–14 days after application.

Description

Skin staples are quick and easy to remove.

An appropriate skin staple remover must be used to remove skin staples 10–14 days after application.

To remove skin staples, grasp the staple remover, place the teeth of the staple remover around the staple, and press the handles of the remover together. This will cause the staple ends to open and lift out of the skin. Pull the staple away from the healed incision and dispose of it appropriately.

PATTERN NARRATION: SKIN STAPLES—REMOVAL

Voice Over: A skin staple remover is used to remove skin staples approximately 10 to 14 days after application.

The staple remover shown in this demonstration has two teeth that slide under the staple and one tooth that is placed over the top of the staple.

Place the teeth of the staple remover around the staple.

Press the handles of the staple remover together, lifting the staple ends out of the skin. Pull the staple away from the incision and dispose of it appropriately.

pattern or technique: suture ligation

Principal Use

Suture ligation is used to prevent the flow of blood to or from an organ or vessel.

Description

Hemostasis is defined as stopping the flow of blood to or from an area in which blood flow is not desired. It is critical to properly ligate, or tie off, blood vessels to ensure hemostasis because significant loss of blood can be fatal.

Blood vessels are often surrounded by connective tissue. A thumb forceps can be used to gently tease away the connective tissue. In most applications, a blood vessel leading to or from an organ will be ligated by allowing enough space between the suture knots to cut the vessel in two locations.

To begin ligating a vessel, tease away the connective tissue and then position the suture material around the vessel near one end. Leave a short end of suture, approximately 1 inch in length, on one side of the vessel and a long end on the other side of the vessel.

Place the needle holder between the two strands of suture. Wrap the long end of the suture around the tip of the needle holder once. Turn the needle holder toward the short end of the suture and grasp the short suture end. Pull the strands with equal tension in a horizontal motion, sliding the long end of the suture over the tip of the needle holder to the opposite side of the wound. The surgeon's hands will cross over each other as the strands are pulled; the short suture end will now be on the opposite side of the wound. Continue to pull horizontally on the ends until the suture material cinches securely around the vessel. This is the first throw.

For the second throw, place the needle holder between the two strands of suture and wrap the long end once around the tip of the needle holder. Turn the needle holder toward the short end of the suture and grasp the suture end. Pull the suture ends apart with equal tension in a horizontal motion, sliding the long end of the suture over the tip of the needle holder to the opposite side of the wound. Failure to alternate direction of the strands with each throw will result in a granny knot.

Each suture knot is made up of a minimum of 3 or 4 throws. To create the third throw, repeat the steps previously described by placing the needle holder between the long and short strands of suture. Wrap the long end of suture once around the tip of the needle holder, grasp the short strand with the needle holder, and pull the suture over the tip of the needle holder to the opposite side of the wound. Repeat the process for the fourth and any additional throws, remembering to alternate the direction of the suture strands with each throw, and pull the strands with equal tension in a horizontal motion. These additional throws will tighten and secure the knot.

After placing the desired number of throws, cut both strands of suture above the knot.

Repeat the process for the second ligation, making sure the suture knot is adequately spaced from the first ligation. Cut the vessel between the two knots and check for bleeding from the ligatures before proceeding.

PATTERN NARRATION: SUTURE LIGATION

Voice Over: When removing an organ, hemostasis must be ensured prior to cutting the attached blood vessels. Suture is often used to ligate a blood vessel, stopping the flow of blood.

In this video, the surgeon is preparing to remove the spleen.

After identifying the blood vessel, grasp the suture with a forceps and pull it around the vessel, leaving a short tab.

Use a suture knot to tie off the blood vessel.

Place the needle holder between the two strands, wrap the suture once around the needle holder, grasp the short end, and pull the suture strands horizontally to the opposite sides. This is the first throw.

For the second throw, wrap the suture once around the needle holder, grasp the short end, and pull the strands horizontally to the opposite sides.

Repeat the process for the third and any additional throws, remembering to alternate direction of the suture strands with each throw.

Cut both strands of suture material above the knot.

Tie a second suture knot closer to the organ before cutting the blood vessel.

Leave enough space between the two suture knots to cut the vessel with an iris scissors.

suture patterns: handouts

square knot/suture knot

Principal use

A **square knot** is used to prohibit or eliminate slipping; a **suture knot** is used to secure a knot in a suture pattern.

Description

The ability to tie a proper suture knot is essential for successfully completing any suture pattern. An improperly tied square knot may result in what is called a granny knot. A granny knot can easily slip and release when under tension and cause a suture pattern to come undone. A complete **square knot** equals 2 throws of suture. It takes a minimum of 2 throws to secure a square knot. A complete **suture knot** is a square knot with the addition of 1 or 2 throws, equaling a minimum of 3 or 4 throws. More throws may be necessary depending on the suture material's coefficient of friction or if the wound is under tension.

Procedure

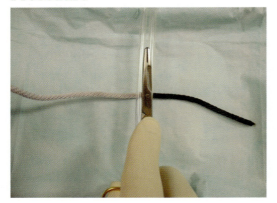

The bicolor string represents the suture material; black is the short strand, and white is the long strand. The plastic tubing represents the incision line. Begin by placing the needle holder over the incision line, between the two strands.

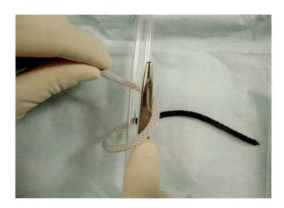

Wrap the long strand once around the end of the needle holder.

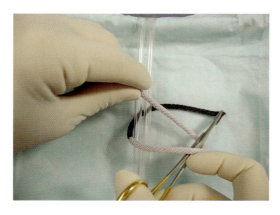

Reach and grasp the short strand with the needle holder.

suture patterns: handouts 59

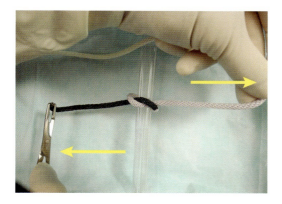

Pull the strands with equal tension in a horizontal motion, bringing the wound edges together until they slightly touch. The surgeon's hand will cross as the short end is pulled to the opposite side of the incision. This is called the first throw.

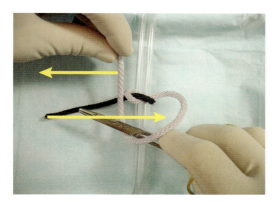

To complete the square knot, loop the long strand around the needle holder, grasp the short strand, and pull the strands in opposite directions as indicated by the arrows. At this point, a square knot has been formed. This is the second throw.

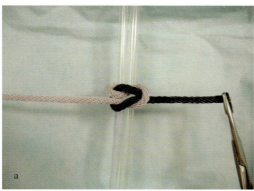
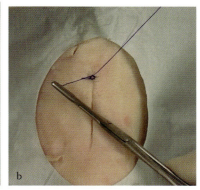

Shown here are the distinct loops of a square knot (a). A square knot equals 2 throws. To create a suture knot (b), a minimum of 3 throws is required. A suture knot is the starting point of most suture patterns. The ability to tie a proper suture knot is essential for successfully completing any suture pattern.

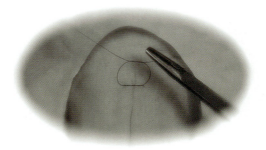

surgeon's knot

Principal Use

The surgeon's knot is used to maintain apposition of the wound edges. It is often used if the wound is under tension or a monofilament suture with a low coefficient of friction is used.

Description

A surgeon's knot is a binding knot used to prevent the first throw from becoming loose. It requires the surgeon to wrap the suture twice around the needle holder, creating increased frictional forces to help the suture stay in place.

Procedure

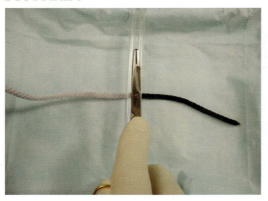

The bicolor string represents the suture material; black is the short strand, and white is the long strand. The plastic tubing represents the incision line. Begin by placing the needle holder over the incision line, between the two strands.

suture patterns: handouts 61

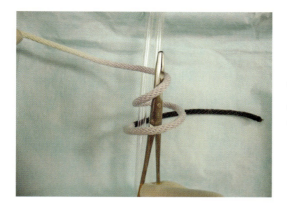

Wrap the long strand twice around the end of the needle holder.

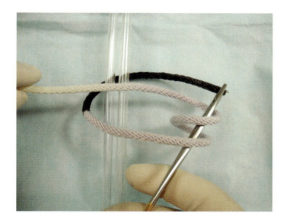

Reach and grasp the short strand with the needle holder.

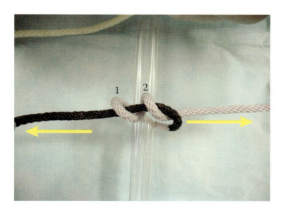

Pull the strands with equal tension in a horizontal motion, bringing the wound edges together until they slightly touch. This is called the first throw. Notice the two loops of the white strand.

A locking throw must be added to complete the surgeon's knot. Loop the long strand around the needle holder, grasp the short strand, and pull the strands as indicated by the arrows. This is the second throw.

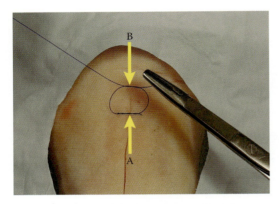

Shown here is the first throw of a surgeon's knot (A) and the second throw (B) before the strands are pulled tight.

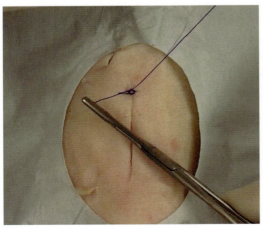

At least 1 additional throw, equaling a minimum of 3 throws, is required to complete the surgeon's knot, depending on the type of suture material used or if the wound is under tension.

suture patterns: handouts 63

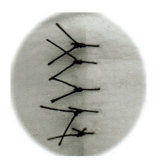

simple interrupted

Principal Use

The basic interrupted suture pattern is used to securely close a wound with accuracy of tissue apposition. It can be used in skin, muscle, organs, vessels, nerves, or fascia.

Description

The simple interrupted suture pattern is easy to place, has good tensile strength, and has less potential for wound edema. This is a secure pattern that allows the surgeon to make adjustments as needed to align the wound edges while suturing. The simple interrupted is a commonly used suture pattern. The surgeon's ability to tie a proper suture knot is essential for successfully completing the simple interrupted suture pattern.

Procedure

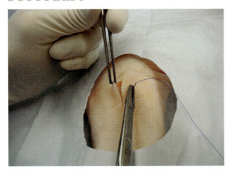

Grasp the skin, evert the wound edge with a forceps, and place the needle perpendicular to the skin. Pass the needle through the skin, following the curve of the needle. Grasp the skin on the opposite side of the wound, evert the edge, and place the needle perpendicular to the inside of the skin. Pass the needle through, following the curve of the needle. The suture should be equidistant from the wound edge on both sides of the incision.

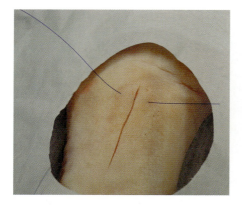

Pull the suture material through the skin, leaving a short end approximately 1 inch in length.

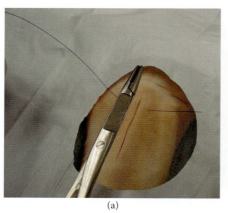
(a)

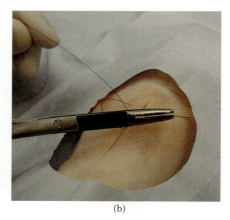
(b)

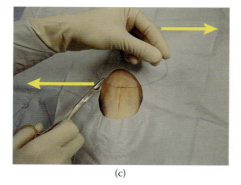
(c)

Place the needle holder between the 2 strands of suture (a). Wrap the long end of the suture around the tip of the needle holder, turn the needle holder toward the short end, and grasp the short end (b). Pull the strands with equal tension in a horizontal motion, sliding the long end over the tip of the needle holder to the opposite side of the wound. The surgeon's hands will cross over each other as the strands are pulled. Bring the wound edges together until they touch, taking care not to evert the wound edges (c). This is the first throw.

suture patterns: handouts 65

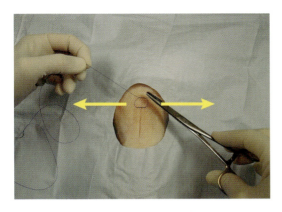

For the second throw, place the needle holder between the two strands of suture and wrap the long end once around the tip of the needle holder. Grasp the short end of the suture and pull the suture ends apart with equal tension in a horizontal motion, sliding the long end of the suture over the tip of the needle holder to the opposite side of the wound. This is the second throw (4a). Failure to alternate direction of the strands with each throw will result in a granny knot.

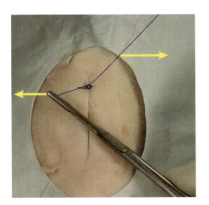

A suture knot equals a minimum of 3 throws. Complete the third and any additional throws, remembering to alternate direction of the suture ends with each throw.

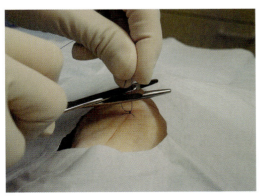

After placing the desired number of throws, cut both strands of suture above the knot, leaving the tabs long enough to grasp easily if the suture is removed.

simple continuous

Principal Use

The simple continuous pattern is useful when quick closure is desired, mainly in long wounds in which tension has been reduced and approximation of the wound edges is acceptable. It may be advantageous in wounds requiring a more airtight or watertight closure.

Description

The simple continuous suture pattern is easy to place and allows for fast wound closure. The simple continuous uses less material, and fewer knots are tied with this pattern, generally resulting in less scarring. However, it offers less security because failure of either knot may result in failure of the entire suture pattern.

Procedure

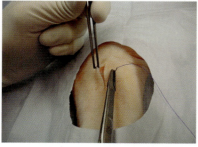

Grasp the skin as shown at left, evert the wound edge with a forceps, and place the needle perpendicular to the skin. Pass the needle through the skin, following the curve of the needle. Grasp the skin on the opposite side of the wound, evert the edge, and place the needle perpendicular to the inside of the skin. Pass the needle through, following the curve of the needle. The suture should be equidistant from the wound edge on both sides of the incision.

suture patterns: handouts

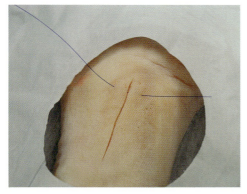

Pull the suture material through the skin leaving a short end approximately 1 inch in length.

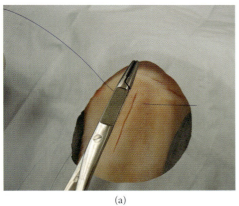
(a)

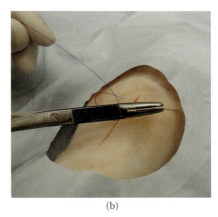
(b)

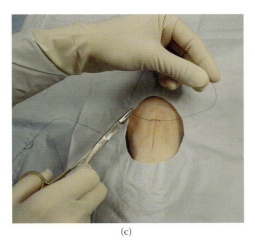
(c)

Place the needle holder between the two strands of suture (a). Wrap the long end of the suture around the tip of the needle holder, turn the needle holder toward the short end, and grasp the short end (b). Pull the strands with equal tension in a horizontal motion, sliding the long end over the tip of the needle holder to the opposite side of the wound. The surgeon's hands will cross over each other as the strands are pulled. Bring the wound edges together until they touch, taking care not to evert the skin edges (c). This is the first throw.

68   suturing principles and techniques in laboratory animal surgery

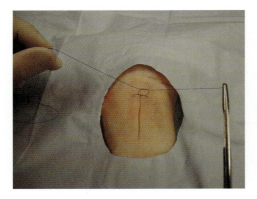

For the second, third, and any additional throws, repeat as previously described by wrapping the suture around the needle holder and pulling the strands with equal tension to the opposite side of the incision.

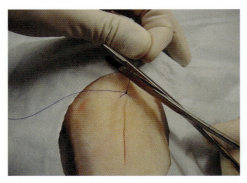

Cut only the short end of the suture material, leaving the long strand to continue the running pattern.

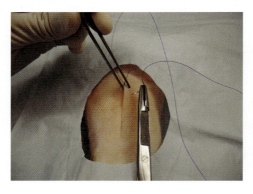

Start the running pattern by grasping the skin below the initial knot, everting the wound edge, then passing the needle through the skin, following the curve of the needle. The suture should be equidistant from the wound edge on both sides of the incision.

suture patterns: handouts 69

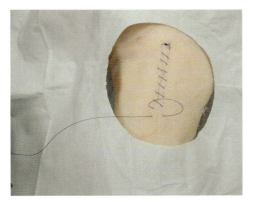

Repeat this step, as previously described, the entire length of the incision. Upon reaching the end of the incision, pull the suture strand through the skin as previously described until a loop is formed approximately 1 inch in length. This loop represents the short end of the suture and will be used to tie the final suture knot.

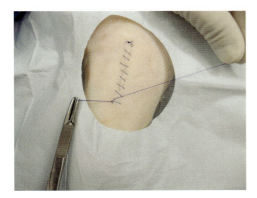

Complete the pattern by tying a suture knot made up of at least 3 or 4 throws.

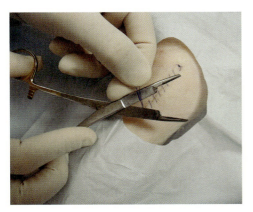

Cut the loop and the long end of the suture, leaving tabs that will be easy to grasp if the suture is removed.

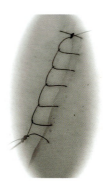

Ford interlocking or lockstitch

Principal Use

The Ford interlocking or lockstitch is useful for closing long skin incisions or wounds under moderate tension.

Description

The lockstitch suture pattern allows for quick closure of a wound and is often more secure than the simple continuous pattern in the case of knot failure. The lockstitch requires more suture material and is more difficult to remove. It is most often used for skin closure in large animals.

Procedure

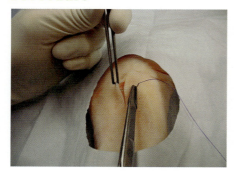

Grasp the skin, evert the wound edge with a forceps, and place the needle perpendicular to the skin. Pass the needle through the skin, following the curve of the needle. Grasp the skin on the opposite side of the wound, evert the edge, and place the needle perpendicular to the inside of the skin. Pass the needle through, following the curve of the needle. The suture should be equidistant from the wound edge on both sides of the incision.

suture patterns: handouts 71

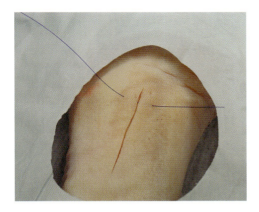

Pull the suture material through, leaving a short end that is approximately 1 inch in length.

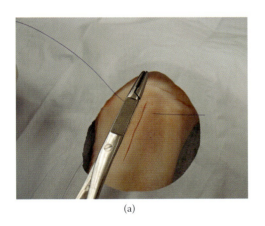
(a)

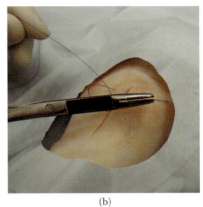
(b)

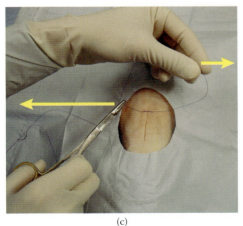
(c)

Place the needle holder between the two strands of suture (a). Wrap the long end of the suture around the tip of the needle holder, turn the needle holder toward the short end, and grasp the short end (b). Pull the strands with equal tension in a horizontal motion, sliding the long end over the tip of the needle holder to the opposite side of the wound. Bring the wound edges together until they touch, taking care not to evert the skin edges (c). This is the first throw.

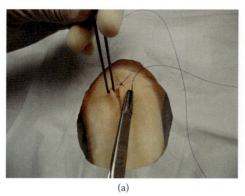
(a)

For the second, third, and any additional throws, repeat as described by wrapping the suture around the needle holder and pulling the strands with equal tension to the opposite side of the incision. Cut only the short end of the suture material, leaving the long strand to continue the running pattern. Start the running pattern by grasping the skin below the initial knot, everting the wound edge, then passing the needle through the skin, following the curve of the needle (a).

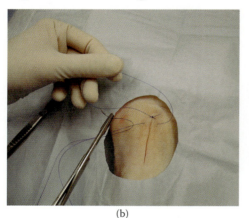
(b)

The suture should be equidistant from the wound edge on both sides of the incision. The needle is now passed through the loop of the preceding stitch (b).

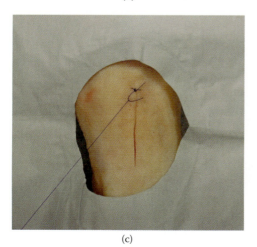
(c)

Pull the suture through and repeat this step the entire length of the incision (c); bring the wound edges together just until they touch, taking care not to evert the skin edges. Repeat this step the entire length of the incision.

suture patterns: handouts 73

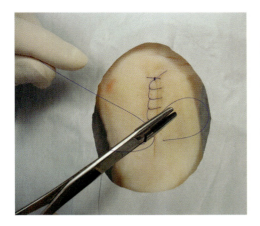

Upon reaching the end of the incision, pull the suture through the skin as previously described until a loop is formed approximately 1 inch in length. This loop will represent the short end of the suture and will be used to tie the final suture knot.

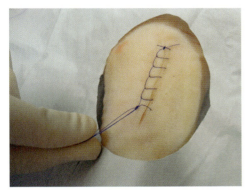

Continue to tie a suture knot with at least 3 or 4 throws. Cut the loop and long end of the suture, leaving tabs long enough to grasp if the suture will be removed.

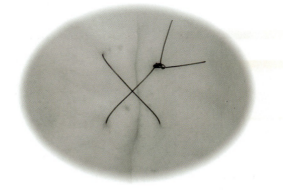

cruciate

Principal Use

The cruciate pattern is primarily used for closing skin wounds under tension.

Description

The cruciate suture pattern is quick and easy to place. It crosses over itself, allowing for a strong closure that is ideal for skin wounds under tension.

Procedure

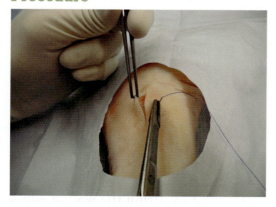

Grasp the skin, evert the wound edge with a forceps, and place the needle perpendicular to the skin. Pass the needle through, following the curve of the needle. Grasp the skin on the opposite side of the wound, evert the edge, and place the needle perpendicular to the inside of the skin. Pass the needle through, following the curve of the needle. The suture should be equidistant from the wound edge on both sides of the incision.

suture patterns: handouts 75

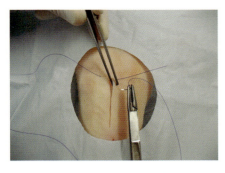

Pull the suture through, leaving a short end approximately 1 inch in length. Bring the suture across the wound, grasp and evert the skin below the initial needle insertion, and place the needle perpendicular to the skin. Pass the needle through, following the curve of the needle. Grasp the skin on the opposite side of the wound, evert the edge, and place the needle perpendicular to the inside of the skin. Pass the needle through, following the curve of the needle. The suture should be equidistant from the wound edge on both sides of the incision.

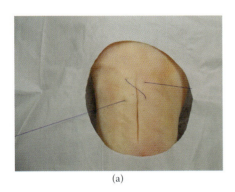

(a)

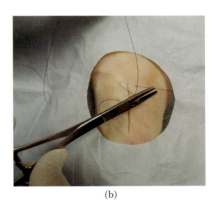

(b)

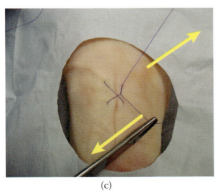

(c)

Pull the long end of the suture through the skin until the wound edges come together, taking care not to invert the edges. This will form a diagonal pattern (a). Place the needle holder between the two strands of suture. Wrap the long end of the suture around the tip of the needle holder, turn the needle holder toward the short end, and grasp the short end (b). Pull the strands with equal tension in a horizontal motion, sliding the long end over the tip of the needle holder to the opposite side of the wound (c). Bring the wound edges together until they touch, taking care not to invert the skin edges. This is the first throw.

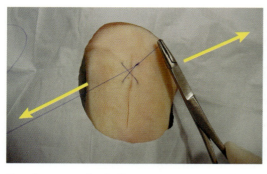

For the second throw, place the needle holder between the two strands of suture, and wrap the long end once around the tip of the needle holder. Grasp the short end of suture and pull the suture ends apart with equal tension in a horizontal motion, sliding the long end of the suture over the tip of the needle holder to the opposite side of the wound. This is the second throw.

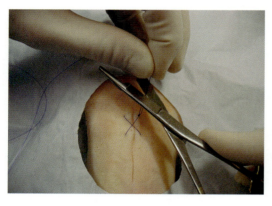

Each suture knot is made up of 3 or 4 throws. Repeat the steps previously described for the remaining throws, remembering to alternate the direction of the suture strands with each throw. Failure to alternate direction of the strands with each throw will result in a granny knot. After placing the desired number of throws, cut both strands of suture above the knot, leaving the tabs long enough to easily grasp if the suture will be removed.

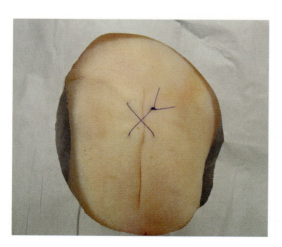

The cruciate is a stronger closure than a simple interrupted and is a good choice in areas under tension. The best apposition is achieved when the corners form a square.

suture patterns: handouts 77

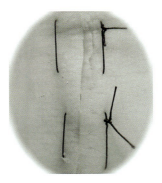

interrupted horizontal mattress

Principal Use

The interrupted horizontal mattress is used for skin, muscle, or tendons that are under high tension.

Description

The interrupted horizontal mattress suture pattern provides good strength and wound eversion. It is generally used in areas that are under high tension. It also functions as a "stay stitch" to temporarily approximate the wound edge while allowing placement of a simple interrupted or simple continuous pattern. Less scarring occurs with this pattern because fewer knots are tied, although the number of needle insertions remains the same.

Procedure

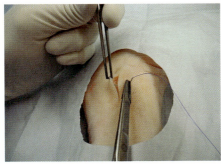

Grasp the skin, evert the wound edge with a forceps, and place the needle perpendicular to the skin. Pass the needle through, following the curve of the needle. Grasp the skin on the opposite side of the wound, evert the edge, and place the needle perpendicular to the inside of the skin. Pass the needle through, following the curve of the needle. The suture should be equidistant from the wound edge on both sides of the incision.

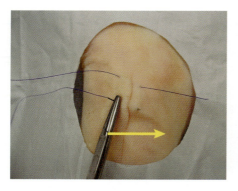

Pull the suture through, leaving a short end that is approximately 1 inch in length. On the same side of the incision, stabilize the skin with a forceps below the location where the needle just exited. Place the needle perpendicular to the skin, insert the needle through the skin, and exit perpendicular to the skin, on the opposite side of the wound. The suture should be equidistant from the wound edge on both sides of the incision. Grasp the needle and pull the suture material through the skin. Both the long and short ends of the suture will now be on the same side of the wound.

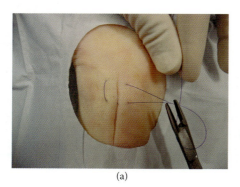

(a)

Place the needle holder between the two strands of suture. Wrap the long end of the suture around the tip of the needle holder, turn the needle holder toward the short end, and grasp the short end (a).

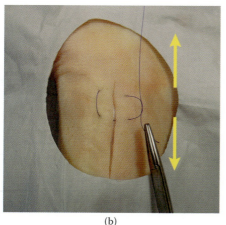

(b)

Pull the strands apart with equal tension in a horizontal motion as indicated by the arrows (b), sliding the long end of the suture over the tip of the needle holder (b). The short suture end will now be on the opposite side. This is the first throw.

suture patterns: handouts

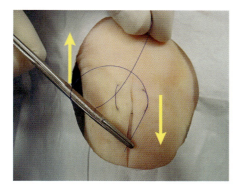

For the second throw, place the needle holder between the two strands of suture and wrap the long end once around the tip of the needle holder. Grasp the short end of suture and pull the suture ends apart with equal tension in a horizontal motion, sliding the long end of the suture over the tip of the needle holder to the opposite side of the wound. This is the second throw.

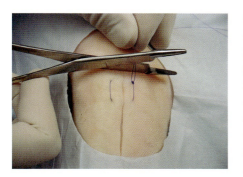

Each suture knot is made up of 3 or 4 throws. Repeat the steps previously described for the remaining throws, remembering to alternate the direction of the suture strands with each throw. Failure to alternate direction of the strands with each throw will result in a granny knot. After placing the desired number of throws, cut both strands of suture above the knot.

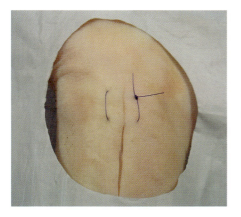

Leave the tabs long enough to grasp easily if the suture will be removed.

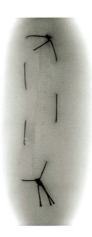

continuous horizontal mattress

Principal Use

The continuous horizontal mattress is used for skin, muscle, or tendons that are under high tension.

Description

The continuous horizontal mattress suture pattern provides good strength and wound eversion. It is generally used in areas that are under high tension. Less scarring occurs with this pattern because fewer knots are tied, although the number of needle insertions remains the same. The continuous horizontal mattress suture pattern allows for faster wound closure than the interrupted horizontal mattress suture pattern; however, failure of either knot may result in failure of the entire suture pattern.

Procedure

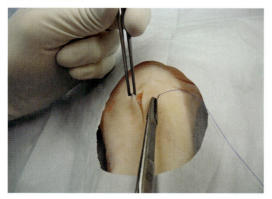

Grasp the skin, evert the wound edge with a forceps, and place the needle perpendicular to the skin. Pass the needle through, following the curve of the needle. Grasp the skin on the opposite side of the wound, evert the edge, and place the needle perpendicular to the inside of the skin. Pass the needle through, following the curve of the needle. The suture should be equidistant from the wound edge on both sides of the incision. Pull the suture through, leaving a short end that is approximately 1 inch in length.

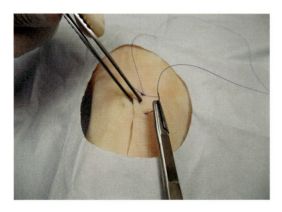

After completing the suture knot with a minimum of 3 or 4 throws, leave the long end intact and cut only the short end. The long end will be used to create a running horizontal mattress pattern the entire length of the incision. Start the running pattern by grasping the skin below the knot, evert the skin, and pass the needle through, doing the same on the opposite side. Pull the suture material through until the wound edges touch. The suture should be equidistant from the wound edge on both sides of the incision.

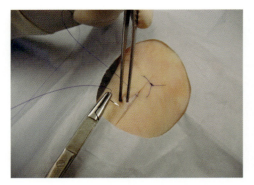

Stabilize the skin with a forceps below the location where the needle just exited. Place the needle perpendicular to the skin, insert the needle through the skin, and exit perpendicular to the skin on the opposite side of the wound.

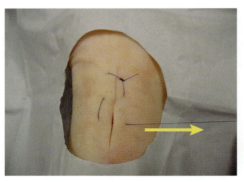

Grasp the needle with the needle holder and pull the suture material through the skin. The long end of suture will now be on the same side of the wound as the initial knot. The suture should be equidistant from the wound edge on both sides of the incision.

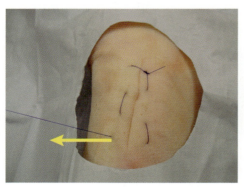

Repeat these steps, alternating sides from right to left, then left to right, the entire length of the incision.

suture patterns: handouts 83

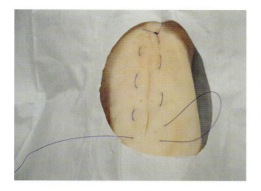

Upon reaching the end of the incision, pull the suture strand through the skin until a loop is formed approximately 1 inch in length.

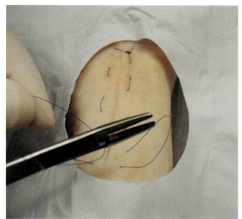

This loop represents the short end of the suture and will be used to tie the final suture knot. Continue to tie a suture knot with at least 3 or 4 throws.

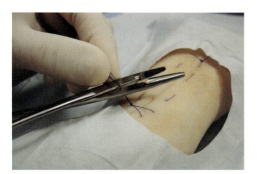

Cut the loop and the long end of the suture, leaving tabs that will be easy to grasp if the suture will be removed.

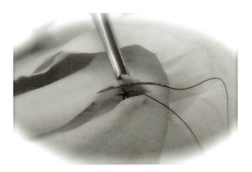

subcuticular

Principal Use

The subcuticular suture pattern is used on areas where a visible suture line is not desirable. This pattern is advantageous when working with species that have a tendency to pick at exposed suture material, such as primates.

Description

The subcuticular suture pattern can be a difficult pattern to place, but when applied properly provides good skin apposition with little to no scarring. The entire pattern is applied into the dermal or subcuticular layer just below the epidermis, allowing it to be hidden beneath the skin. This pattern can be helpful when working with species that have a tendency to pick at exposed suture material, such as primates. The subcuticular pattern is weaker than skin sutures; therefore, some skin closures may be added if desired.

suture patterns: handouts 85

Procedure

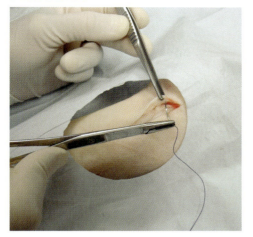

A subcuticular suture pattern begins with an anchor knot buried beneath the skin. Starting at one end of the incision, insert the needle below the epidermis, deep into the dermal or subcuticular layer. Exit vertically through the dermal layer, just below the epidermis.

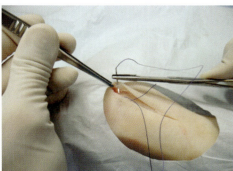

Pull the suture through, leaving a short end that is approximately 1 to 2 inches in length. Then, on the opposite side of the incision, insert the needle into the dermal layer, just below the epidermis, and exit vertically out the deeper dermal or subcuticular layer.

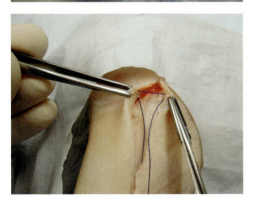

Here is a view of the suture placed into the subcuticular layer prior to tying the knot.

86   suturing principles and techniques in laboratory animal surgery

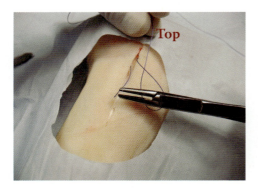

Tie the anchor knot by placing the needle holder between the two strands of suture. Wrap the long end of the suture around the tip of the needle holder, turn the needle holder toward the short end, and grasp the short end.

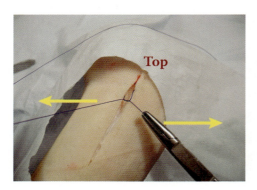

Pull the strands with equal tension in a horizontal motion, sliding the long end over the tip of the needle holder to the opposite side of the wound. Bring the wound edges together until they touch, taking care not to evert the skin edges. This is the first throw. For the second, third, and any additional throws, repeat as described by wrapping the suture around the needle holder and pulling the strands with equal tension to the opposite side of the incision.

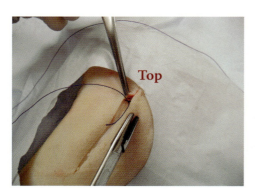

The completed knot will be buried under the skin. Cut only the short end of the suture material, leaving the long strand to continue the running pattern.

suture patterns: handouts 87

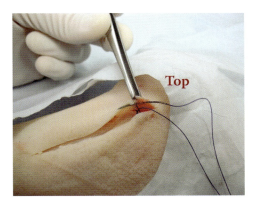

Start the running pattern by grasping the skin at the beginning (top) of the incision and everting the wound edge. Place the needle under the epidermis into the dermal layer and pass the needle horizontally through the dermis.

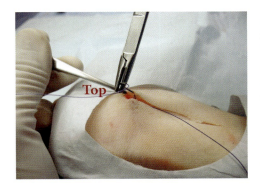

Bring the wound edges together just until they touch, taking care not to invert the skin edges. Grasp the skin on the opposite side, evert the edge, again place the needle under the epidermis into the dermal layer, and pass the needle horizontally through the dermis, bringing the wound edges together.

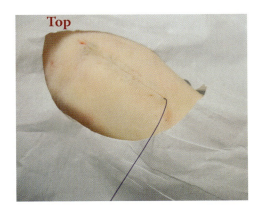

Continue passing the needle horizontally through the dermal layer, alternating sides the entire length of the incision.

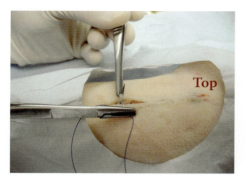

Upon reaching the end of the incision, complete the suture pattern with a buried knot. Grasp the skin toward the end of the incision and evert the wound edge. Insert the needle below the epidermis, deep into the dermal or subcuticular layer. Exit vertically through the dermal layer, just below the epidermis.

Pull the suture through until a loop is formed approximately 1 inch in length. This loop will represent the short end of the suture and will be used to tie the final suture knot.

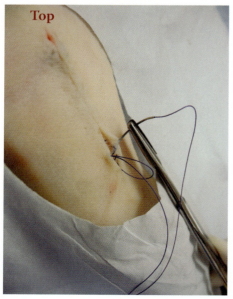

Then, on the opposite side of the incision, insert the needle into the dermal layer, just below the epidermis, and exit vertically out the deeper dermal or subcuticular layer.

suture patterns: handouts 89

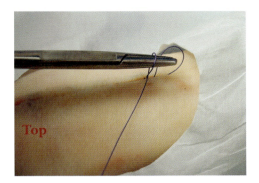

Place the needle holder between the long strand of suture and the short loop. Wrap the long end of the suture once around the tip of the needle holder. Turn the needle holder toward the loop and grab the suture at the apex of the loop.

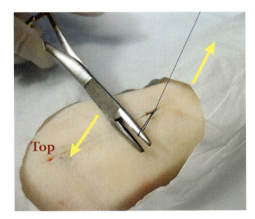

Pull the strands with equal tension in a horizontal motion, sliding the long end of the suture over the tip of the needle holder to the opposite side of the wound. The short loop will now be on the opposite side of the wound. This is the first throw.

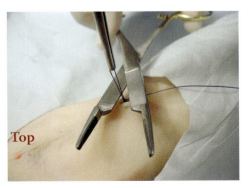

Continue to tie a suture knot with at least 3 or 4 throws. Leave the long end intact and cut only the short loop, just above the knot.

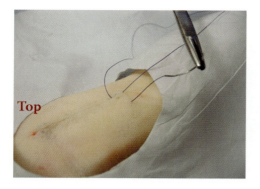

Insert the needle vertically back into the incision, just behind the final buried knot. Turn the needle away from the incision line and exit the skin.

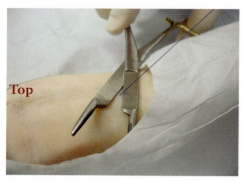

Apply tension to the suture, pulling the buried knot deeper into the tissue. Cut the suture strand flush with the skin layer.

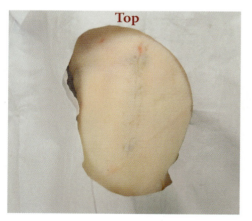

Shown here is a picture of a completed subcuticular suture pattern.

suture patterns: handouts 91

skin staples

Principal Use

Skin staples are primarily used for skin closure especially in wounds under tension. Staples may be used in place of suture to prevent an animal from chewing through the suture line.

Description

Skin staples are quick and easy to apply when closing skin wounds under tension. They are often used instead of suture to prevent an animal from chewing through the suture line and opening the wound.

Procedure

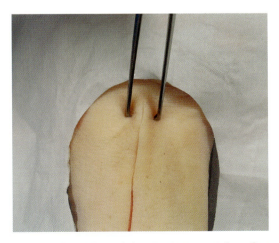

Grasp the skin on both sides of the incision with a forceps and evert the skin layers, keeping the wound edges together.

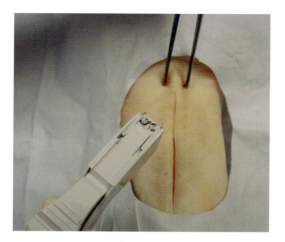

Grasp the skin stapler with the dominant hand.

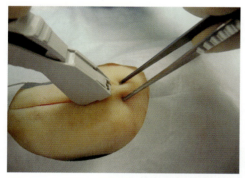

Place the tip of the stapler against the everted skin, just below the forceps.

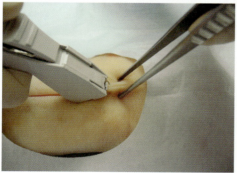

Apply the staple by pressing the applicator together, causing the staple to insert into the skin layer.

suture patterns: handouts 93

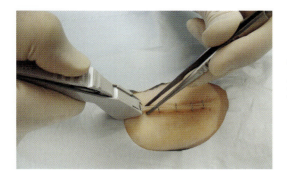

The staples should be placed equidistant from the wound edge on both sides of the incision.

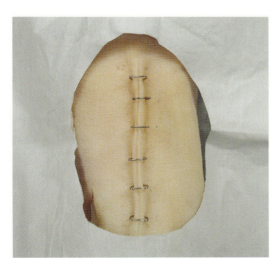

Depending on the type of staple and applicator used, the staple may not be flush with the skin and may stand up off the skin layer. This allows room for swelling, which is often associated with tissue trauma. Staples are removed with a specified staple remover 10–14 days after application.

skin staples: removal

Principal Use

A staple remover is used to remove skin staples from an incision 10–14 days after application.

Description

Skin staples are quick and easy to remove.

Procedure

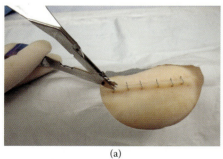

(a)

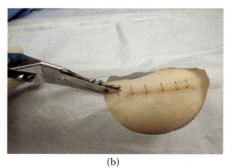

(b)

(c)

Grasp the staple remover, place the teeth of the remover around the staple (a), and press the handles of the remover together (b). This will cause the staple ends to open and lift out of the skin. Pull the staple away from the healed incision (c) and dispose of appropriately.

study breaks: DVD narration

Study Break 1: "Needle Placement on Needle Holder"

After opening the suture package, grasp the needle with the needle holder and remove the needle and suture material completely from the package.

Grasp the suture near the swaged end of the needle to stabilize and reposition the needle on the needle holder.

Grasp the needle two-thirds of the way down, away from the tip, and secure it in the needle holder by locking the ratchet lock. If grasped correctly, the needle will be perpendicular to the needle holder.

Hold the needle holder out in front of you as if you were shaking hands; the needle should be smiling back at you.

Study Break 2: "Attach and Remove Blade Safely (Nonsterile)"

After opening the blade package, look at the base of the blade for the size number; it is only on one side of the blade. Make sure the side with the number is facing up.

Using a needle holder, grasp the blade above its opening on the side opposite the cutting edge. Lock the ratchet lock on the needle holder to secure the blade and remove the blade from the packaging.

The surgeon's hand needs to be appropriately placed on the needle holder with the index finger placed near or on the box lock.

Notice the angle at the base of the blade. This angle needs to correspond to the angle on the blade handle as shown.

Slide the blade opening over the tip of the blade handle and use the index finger to push the blade onto the handle until it clicks in place. Remove the needle holder and inspect the blade to ensure it is on the handle correctly.

To remove the blade, grasp the base of the blade with the tip of the needle holder; lock the ratchet lock.

Rotate the needle holder to lift the base of the blade off the surface of the handle, using the thumb of the nondominant hand; push the needle holder up, allowing the blade to be removed from the handle in a controlled manner.

Discard the blade in an appropriate sharps container.

Study Break 3: "Make an Incision"

Place the thumb and index finger of the nondominant hand on the tissue surface to stabilize and add tension to the skin.

Place the belly of the blade on the tissue surface, apply downward pressure with the index finger, and draw the blade across the tissue surface in one, smooth, complete motion.

If performed correctly, the skin layer will part as shown.

suture considerations for exotics and other species

introduction

Exotics and other species (e.g., rabbits, ferrets, fish, birds, amphibians, etc.) are commonly used in research. Differences in species anatomical, physiological, or tissue characteristics may alter the surgeon's technique concerning how the wounds are closed and the type of suture material used.

Information addressing wound closure techniques is becoming readily available with the increase in associations, journals, and conferences focused on exotic animals.

Multiple studies have shown that there is not one specific suture material that is appropriate for the variety of applications found in veterinary medicine, and this especially holds true with exotic animals. At a minimum, suture material needs to be chosen based on the procedure, the material's physical properties, and the effect it has on wound healing. As a general rule, it is best to choose the smallest-diameter suture possible that will still retain its strength long enough for the tissues to heal.

The following summaries are intended to provide general guidelines, based on current literature, for choosing appropriate suture material for exotics and other species. A laboratory animal veterinarian should be contacted to discuss the specifics of a procedure to ensure the proper suture choice is made.

rabbit

1. Studies indicated suturing the urogenital tract or bladder with a braided, absorbable material causes diminished tissue response when compared to nylon, nonabsorbable material.[1,2]
2. Skin incisions are usually closed using absorbable suture material in a subcuticular pattern. Specifically, a monofilament suture with a swaged needle works best as it will reduce tissue trauma.[3]
3. Skin staples and tissue glue are well tolerated by the rabbit and generally do not cause skin irritation.

ferret

1. In regard to ferret spays, the muscle layer can be closed with stainless steel. These sutures have minimum tissue reaction and hold together well. Suture staples can be used for closing the skin, but an absorbable monofilament applied using a subcuticular pattern is also acceptable.[4]
2. If sutures are properly placed and not overly tight, an absorbable suture can be used for the muscle layer and a nonabsorbable suture can be used for the skin. The ferret will generally leave the sutures alone if they are not too tight.

birds

1. Birds seem to have an inflammatory response to most suture materials except stainless steel. Unfortunately, using stainless steel can lead to hematomas and seromas. An absorbable monofilament may produce the smallest inflammatory response.[5]
2. A braided absorbable material has been mentioned for use in bird skin because the soft suture is less irritating and is quickly absorbed.[6]

reptiles and amphibians

1. When closing the muscle layer, the suture material chosen needs to be absorbed as quickly as possible while maintaining strength long enough for the tissue to heal. A braided absorbable is a good choice in this case.[7]
2. The use of absorbable monofilament is recommended for skin closure as studies showed it has the least histological reaction.[8,9]

fish

1. The use of absorbable monofilament of adequate size is recommended for use in fish. A direct comparison between various types of monofilament materials has not been performed to date.[6,10]

works cited

1. Gomel V, McComb P, Boer-Meisel M. Histologic reactions to polyglactin 910, polyethylene, and nylon microsuture. *J Reprod Med* 25:56–59, 1980.
2. Beauchamp PJ, Guzick DS, Held B, et al. Histologic response to microsuture materials. *J Reprod Med* 33:615–623, 1988.
3. Longley L. *Saunders Solutions in Veterinary Practice: Small Animal Exotic Pet Medicine.* Saunders, Philadelphia, 2010.
4. Client Education Communication, Long Beach Animal Hospital 3816. Long Beach, CA, 1998–2013. www.lbah.com/word/spay-ferret
5. Bennett RA, Yaeger MJ, Trapp A, et al. Histologic evaluation of the tissue reaction to five suture materials in the body wall of rock doves (*Columba livia*). *J Avian Med Surg* 11:175–182, 1997.
6. McFadden MS. Suture materials and suture selection for use in exotic pet surgical procedures. *J Exotic Pet Med* 20:173–181, 2011.

7. McFadden MS, Bennett RA, Kinsel MJ, Mitchell MA. Evaluation of the histologic reactions to commonly used suture materials in the skin and musculature of ball pythons. *Am J Vet Res* 72:1397–1406, 2011.
8. Tuttle AD, Law JM, Harms CA, et al. Evaluation of the gross and histologic reactions to five commonly used suture materials in the skin of the African clawed frog (*Xenopus laevis*). *J Am Assoc Lab Anim Sci* 45:22–26, 2006.
9. Green SL. *The Laboratory* Xenopus *sp.* CRC Press, Taylor and Francis Group, Boca Raton, FL, 2010.
10. Roberts HE. Aquatic veterinary services: Surgical management of pet fish. AVMA Convention CE Course 2010 Surgery Fish.

glossary

Apex: The top or highest part of the suture loop.

Apposition: The positioning of things or the condition of being side by side or close together.

Burying the knot: When suturing, having the knot of the suture buried beneath the skin surface.

Continuous suture pattern: One in which a continuous, uninterrupted length of material is used.

Deep to superficial: When suturing, going from the deepest skin layers to the most superficial (just under the surface) skin layers.

Dermis: The thick layer of living tissue below the epidermis that forms the true skin; contains the capillaries, nerve endings, sweat glands, hair follicles, and other structures.

Granny knot: An insecure knot often made instead of a square knot.

Ligature: Suture material used to tie or bind something tightly (i.e., blood vessel).

Long strand: When tying a suture knot, the strand of material that has the needle attached.

Low coefficient of friction: Characteristic of suture material that allows strands to slide easily over each other. With sutures, monofilaments have a lower coefficient of friction than a braided material.

Percutaneous: In the skin.

Running pattern: Same as a continuous suture pattern.

Short strand: When tying a suture knot, the end of the material without the needle.

Square knot: A type of secure double knot that is made symmetrically to hold securely.

Strand: A single, thin length of suture material.

Subcutaneous: Under the skin.

Subcuticular: Tissue beneath the skin.

Superficial to deep: When suturing, going from the outer skin layers to the deeper skin layers.

Surgeon's knot: The surgeon's knot is a simple modification to the suture knot. The suture material is wrapped twice around the needle holder on the first throw of the surgeon's knot, creating increased frictional forces to help the suture stay in place.

Suture knot: A suture knot is a square knot with the addition of at least 1 or 2 throws, equaling a minimum of 3 or 4 throws.

Tab (suture): The suture material remaining after cutting the strand on completion of a suture knot.

Throw: An essential component of the suture knot resulting from passing the end of the short suture strand through a loop formed by the long suture strand. A completed suture knot consists of at least 3 to 4 throws.

index

A

Absorbable sutures
   absorption rate, 14
   types of, 15
   uses, 14, 15, 16
Adson Brown forceps, 4–5
Adson forceps, 4–5
Amphibians, wound closure techniques, 99

B

Birds, wound closure techniques, 98
Blades
   attaching to handle, 7, 95–96
   choosing, 6–7
   making incisions, 8, 96
Blood vessels, ligating, 54–55

C

Closed eye needle, 10
Continuous horizontal mattress pattern
   description, 41, 80
   narration, 44–45
   procedure, 41–45, 81–83
   use, 41, 80
Continuous patterns. *See* Continuous horizontal mattress pattern; Lockstitch pattern; Simple continuous pattern; Subcuticular pattern
Cruciate pattern
   description, 35, 74
   narration, 36–37
   procedure, 35–37, 74–76
   use, 35, 74
Curved needles, 11
Cutting edge needles
   conventional, 9, 11–12
   reverse, 9, 11–12
   uses, 9

F

Ferrets, wound closure techniques, 98
Fish, wound closure techniques, 99
Forceps, 4–5
Ford interlocking pattern. *See* Lockstitch pattern

G

Granny knot, 19, 57

H

Hematomas, 98
Hemostasis, 54

I

Incisions, making, 8, 96
Inflammatory response, to stainless steel suture materials, 98
Interrupted horizontal mattress pattern
   description, 38, 77
   narration, 39–40
   procedure, 38–40, 77–79
   use, 38, 77

103

Interrupted suture patterns. *See*
	Cruciate pattern; Interrupted horizontal mattress pattern; Simple interrupted pattern
Iris scissors, 6

L

Ligation, 54–55
Lockstitch pattern
	description, 31, 70
	narration, 33–34
	procedure, 31–34, 70–73
	use, 31, 70

M

Mattress suture patterns. *See* Continuous horizontal mattress pattern; Interrupted horizontal mattress pattern
Mayo Hegar needle holder, 1–3
Mayo scissors, 6
Metzenbaum scissors, 5–6
Monofilament suture
	characteristics of, 14
	uses, 15, 17
Multifilament suture
	characteristics of, 14
	uses, 15, 16

N

Needle holders
	choosing, 2
	holding, 1
	Mayo Hegar, 1–3
	needle placement, 2–4, 95
	Olsen Hegar, 1–3
Needles
	body of, 11
	characteristics, 9
	choosing, 9
	closed eye, 10
	components, 10–12
	curved, 11
	cutting edge, 9, 11–12
	eye, 10–11
	placement on needle holder, 2–4, 95
	point of, 11–12
	straight, 11
	swaged, 10–11
	taper point, 9, 10, 11, 12
Nonabsorbable sutures
	types of, 14, 15
	uses, 15, 16

O

Olsen Hegar needle holder, 1–3
Operating scissors, 1, 5–6
Organ removal, and suture ligation, 54–55

R

Rabbits, wound closure techniques, 98
Reptiles, wound closure techniques, 99
Round needle, 12
Running patterns. *See* Continuous horizontal mattress pattern; Lockstitch pattern; Simple continuous pattern; Subcuticular pattern

S

Scalpel handle
	blade choices, 6–7
	sizes, 6
Scissors
	holding, 1, 6
	types of, 5–6
Seromas, 98
Simple continuous pattern
	description, 27, 66
	narration, 29–30
	procedures, 27–30, 66–69
	use, 27, 66
Simple interrupted pattern
	description, 24, 63
	narration, 25–26
	procedure, 24–26, 63–65
	use, 24, 63
Skin staples. *See* Staples
Square knot
	description, 19, 57
	pattern narration, 21
	procedure, 19–21, 58–59
	use, 19, 57
Stainless steel suture materials, inflammatory response to, 98
Staples
	description, 51, 91
	narration, 52, 53
	procedure, 51–52, 91–93
	removal, 53, 94
	use, 51, 91
Stay stitch, 38, 77
Straight needles, 11
Subcuticular pattern
	description, 46, 84
	narration, 48–50

procedure, 46–50, 84–90
use, 46, 84
Surgeon's knot
  description, 22, 60
  pattern narration, 23
  procedure, 22–23, 60–62
  use, 22, 60
Suture knot
  description, 19, 57
  patter narration, 21
  procedure, 19–21, 58–59
  use, 19, 57
Suture ligation, 54–55
Suture material
  absorbable, 14–15, 16
  choosing, 13, 16–17
  monofilament, 14, 15, 17
  multifilament, 14, 15, 16
  nonabsorbable, 14, 15, 16
  packaging, 16
  sizes, 13
  types of, 14–15

Suture patterns. *See individual patterns*
Swaged needles, 10–11

T

Taper point needles, 9, 10, 11, 12
Tissue forceps
  holding, 5
  types of, 4–5
  uses, 4

W

Wound closure techniques
  amphibians, 99
  birds, 98
  ferrets, 98
  fish, 99
  rabbits, 98
  reptiles, 99

Heterick Memorial Library
Ohio Northern University

AUG 28 2013

DUE	RETURNED	DUE	RETURNED
1.		13.	
2.		14.	
3.		15.	
4.		16.	
5.		17.	
6.		18.	
7.		19.	
8.		20.	
9.		21.	
10.		22.	
11.		23.	
12.		24.	

Withdrawn From
Ohio Northern
University Library

OHIO NORTHERN UNIVERSITY
3 5111 00671 7585

Heterick Memorial Library
Ohio Northern University
Ada, Ohio 45810